War by Agreement

War by Agreement presents a new theory on the ethics of war. Benbaji and Statman argue that wars can be morally justified at both the ad bellum level (the political decision to go to war) and the in bello level (the actual conduct of the military) by accepting a contractarian account of the rules governing war. According to this account, the rules of war are anchored in a mutually beneficial and fair agreement between the relevant players—the purpose of which is to promote peace and to reduce the horrors of war. This account relies on the long social contract tradition and illustrates its fruitfulness in understanding and developing the morality and the law of war.

Yitzhak Benbaji teaches philosophy at Tel-Aviv University Faculty of Law. He previously worked in the Department of Philosophy and in the Faculty of Law at Bar-Ilan University (2002–12). He has been a visiting scholar and professor at the Princeton Institute for Advanced Study, the University of Toronto, Northwestern University, Yale Law School, and the Institute for Advanced Study in Jerusalem. Among his publications are *The View from Within* (2011) and *Reading Walzer* (2013).

Daniel Statman teaches at the philosophy department at the University of Haifa and is former Chair of the Israeli Philosophical Association. He has been a visiting scholar and professor at various institutes, including the Princeton Institute for Advanced Study, the University of Michigan, Oxford University, and the Forschungskolleg Humanwissenschaften in Bad Homburg. He is the author and editor of many books and articles, including *Moral Luck* (1993), *Moral Dilemmas* (1995), *Religion and Morality* (1995), *Virtue Ethics* (1997), and *State and Religion in Israel* (2019).

War by Agreement

A Contractarian Ethics of War

YITZHAK BENBAJI

and

DANIEL STATMAN

Great Clarendon Street, Oxford, OX2 6DP,
United Kingdom

Oxford University Press is a department of the University of Oxford.
It furthers the University's objective of excellence in research, scholarship,
and education by publishing worldwide. Oxford is a registered trade mark of
Oxford University Press in the UK and in certain other countries

© Yitzhak Benbaji and Daniel Statman 2019

The moral rights of the authors have been asserted

First published 2019
First published in paperback 2023

All rights reserved. No part of this publication may be reproduced, stored in
a retrieval system, or transmitted, in any form or by any means, without the
prior permission in writing of Oxford University Press, or as expressly permitted
by law, by licence or under terms agreed with the appropriate reprographics
rights organization. Enquiries concerning reproduction outside the scope of the
above should be sent to the Rights Department, Oxford University Press, at the
address above

You must not circulate this work in any other form
and you must impose this same condition on any acquirer

Published in the United States of America by Oxford University Press
198 Madison Avenue, New York, NY 10016, United States of America

British Library Cataloguing in Publication Data

Data available

Library of Congress Cataloging in Publication Data

Data available

ISBN 978–0–19–957719–4 (Hbk.)
ISBN 978–0–19–891078–7 (Pbk.)

DOI: 10.1093/oso/9780199577194.001.0001

Links to third party websites are provided by Oxford in good faith and
for information only. Oxford disclaims any responsibility for the materials
contained in any third party website referenced in this work.

To Ido and Noa
YB

To Michal
DS

Table of Contents

Preface	ix
Introduction: The Moral Standing of the War Agreement	1
The War Agreement: *Jus ad Bellum*	3
The War Agreement: *Jus in Bello*	4
The Moral Standing of a Sufficiently Good Legal System	6
The Structure of the Book	7
1. The Challenge	9
1.1 Introduction	9
1.2 The Traditional View	10
1.3 Individualism and Continuity	13
1.4 *Individualism* and the Moral Justification to Wage War	15
1.5 *Individualism* and the Moral Rules for Conducting Warfare	21
1.6 Conclusion	35
2. Foundations of a Non-Individualist Morality	37
2.1 Introduction	37
2.2 Social Rules and Morality	38
2.3 The Conditions for the Moral Effectiveness of Social Rules	43
2.4 Social Roles and the Moral Division of Labour	49
2.5 The Conditions for the Moral Effectiveness of Social Roles	55
2.6 A Moral Duty to Fulfil Professional Duties?	59
2.7 Duels, Executions, and the Importance of Decent Regimes	62
2.8 Between Contractarianism and Rule-Consequentialism	66
2.9 Conclusion	68
3. A Contractarian Account of the Crime of Aggression	71
3.1 Introduction	71
3.2 Why the *Ad Bellum* Regime is Mutually Beneficial	72
3.3 Fairness: Minimally Just Symmetrical Anarchy	80
3.4 A Hypothetical: A Story about Divided and Well-Ordered	86
3.5 Wars of Independence	89
3.6 Self-Help vs Global Police or Global Court	92
3.7 Conclusion	95
4. The Aims of Just Wars and *Jus Ex Bello*	98
4.1 Introduction	98
4.2 Just Cause vs Just Aims	99
4.3 The Logic of Uncompromising War	101

4.4 Historical Agreements on Ending Wars	107
4.5 Deterrence and Prevention: The Just Aims of Defensive Wars	110
4.6 *Ad Bellum* vs *Ex Bello* Proportionality	111
4.7 Conclusion	115
5. Contractarianism and the Moral Equality of Combatants	116
5.1 Introduction	116
5.2 *Moral Equality* and the Importance of Obedient Armies	117
5.3 Defending Obedience	120
5.4 Do Combatants Accept Legal Equality?	124
5.5 Contractarianism and Treachery	129
5.6 Responsibility for Killing Just Combatants	130
5.7 Conclusion	131
6. Contractarianism and the *Moral Equality* of Civilians	133
6.1 Introduction	133
6.2 *Mutual Benefit*	135
6.3 *Fairness*	140
6.4 *Collateral Damage*	142
6.5 *In Bello* Necessity and *In Bello* Proportionality	144
6.6 Civilian Acceptance of the War Agreement	149
6.7 *Civilian Immunity* and *Moral Equality* in Asymmetric Wars	152
6.8 Insights from Behavioural Ethics	158
6.9 Conclusion	161
7. When the Agreement Collapses	163
7.1 Introduction	163
7.2 Ruthless Warfare and the War Agreement	165
7.3 Supreme Emergencies	171
7.4 A Contractarian Account of the Exemption	174
7.5 Conclusion	178
8. Concluding Remarks	181
8.1 Overview of the Argument	181
8.2 The Scope and Limitations of Contractarianism	191
8.3 War and Tragedy	193
8.4 Revisionism and Contractarianism	196
8.5 Last Word	199
Bibliography	201
Index	211

Preface

As we write these lines in late 2018, dozens of wars are still raging around the globe, while parallel to these wars a harsh and quite combative debate is being held. To put it perhaps over-dramatically, we could say that a 'war about war' is also in progress. We refer to the debate between a group of philosophers called 'revisionists' and a comparatively few opposing thinkers who seek to defend a traditional understanding of the ethics of war. The revisionists, led by Jeff McMahan, are challenging all fundamental tenets of traditional just war theory, such as the right granted to soldiers on both sides to kill each other and the requirement to maintain a strict distinction in warfare between combatants and noncombatants. In the revisionist view, if killing in war can ever be morally justified, it must be grounded in the same principle that justifies killing outside the context of war; namely, in accordance with the right to self-defence. They argue, however, that this cherished principle falls short of grounding the above tenets.

What would follow if the revisionists were to win this debate? What are the alternatives to traditional just war theory? One would be to give up the attempt to evaluate war in moral terms and adopt instead the message of the old Latin adage that, in times of war, the law is silent; *inter arma silent leges*. But this realist view of war has been subject to strong criticism and does not seem a very promising avenue. A completely different alternative would be to go for pacifism. If the revisionist critique of traditional just war theory is sound, then although killing in war might be permissible, the odds are that it is not. Given the strong presumption against killing human beings, perhaps the only decent conclusion is to advocate restraint from war altogether.

Although at times revisionists seem to be going in this direction, in the end most of them refrain from doing so. Readers coming to the end of some revisionist writings feel as if they are watching the last minutes of a movie with an unexpected twist in the plot. The surprising twist is the revisionist acceptance of the legal rules governing war and their return to the basic teachings of just war theory. Yet from our perspective, after the powerful criticism that revisionists have mounted against these teachings in the lion's share of their writings, this return is impossible. The way back is blocked.

We believe that these developments have resulted in a stalemate. On the one hand, the revisionist criticism of traditional just war theory has made the older view seem ungrounded and, in a sense, naive. On the other hand, revisionists have failed to propose a convincing alternative. The main purpose of the present book is to forge a way through this stalemate. We aim to show that wars can be morally justified at both the *ad bellum* level (the political decision to go to war) and the *in bello* level (its actual conduct by the military). The alternative to pacifism is not crude realism of the 'sometimes there is just no choice but to go to war' form, but rather a rich theory that grounds the principles regulating war in a mutually beneficial and fair agreement between the relevant players. In other words, what we propose is to ground the accepted rules concerning war in a contractarian framework. Hence the title of this book—*War by Agreement*—which, as some readers must realize, is a variation on the title of a key book on moral contractarianism by David Gauthier from 1986, *Morals by Agreement*. While Gauthier aimed to ground all morality in agreement, our own contractarianism is more limited and, at any rate, is applied here mainly to the morality of engaging and harming in war.

*

Our interest in the ethics of war did not emerge from mere philosophical curiosity. It is connected with the fact that Israel, the country in which we both live, has been involved in wars since its inception, and more importantly, that these wars have been a constant object of moral reflection. The ongoing debate in Israel is concerned both with the *jus ad bellum* and with the *jus in bello* levels. Most saliently for us, regarding the 1982 war in Lebanon, the debate turned on whether the very launching of the war was morally justified; namely, whether it satisfied the *ad bellum* conditions. We note the 1982 war in particular, although similar deliberations surrounded other wars in which Israel was involved in 1948, 1956, 1967, and 1973, because we experienced the debates over the 1982 war 'live' while the other armed struggles took place before we were born or before we matured philosophically. Since then, the justice of the military occupation in the West Bank and in the Gaza Strip, together with the asymmetric wars that maintain it, have become a topic of bitter and heated controversy. As happens frequently, the debate among Israelis has turned on the measures and tactics used in the course of the fighting as well; namely, to questions regarding *jus in bello*. As philosophers trained in moral and political philosophy, we naturally found ourselves trying to utilize our philosophical tools to shed light on and find solutions to these controversial issues. Danny's first essay in this field, published twenty years ago, was about *jus in bello* and the (first) Intifada, to be followed several years later by a paper on targeted killing (which was then a new and controversial tactic used by Israel against terror organizations).

The internal debate in Israel about the morality of its wars and of its military operations more generally led, perhaps inevitably, to a debate about the legitimacy of disobeying orders to participate in missions to which soldiers are strongly opposed; for instance, orders to maintain the military occupation of the West Bank and the Gaza Strip and to participate in the military operations involved in the occupation. Some have argued that except in cases of manifestly immoral orders, soldiers are under a duty to obey orders even if they object to the specific actions involved and beyond that—even if they object to the war in general. Others have taken an opposing position and argued that it is the moral duty of soldiers to refuse to participate in what they consider to be unjust wars or unjust missions.

At the time that each of us was starting to develop his own thoughts on the ethics of war, revisionism was spreading swiftly among philosophers and legal theorists. Like most people working in the field, we found ourselves perplexed by the teachings of this new movement. We appreciated the force of its arguments and realized how devastating they were for traditional just war theory, but could see no alternative. For several years, revisionism seemed like the only game in town.

But then, while Danny was still working within the individualist paradigm set by the revisionists (see his papers on 'supreme emergencies' and on the 'success condition'), Yitzhak started moving in a new direction—offering a contractarian basis for traditional just war theory. The idea wasn't altogether new, but in the past it had been used only to explain some limited aspects of the war ethic without providing a theoretical underpinning. Yitzhak undertook the task of its theoretical grounding in a series of articles published over the last several years. Initially, Danny was sceptical about this direction. But in the course of time and following long discussions, he too became convinced that contractarianism provided the best (probably the only) stable basis for the traditional *ad bellum* and *in bello* principles. He became convinced that if anything works (in terms of justifying these principles), it is contractarianism. He also came to see that, as a general view of morality, contractarianism is much more attractive than he had previously thought. The result was that Danny also started to rely on contractarianism in his writings on the ethics of war. It didn't take long for Yitzhak to invite Danny to join him in authoring a full-length manuscript on the contractarian view of war.

As is usually the case with academic writing, the project took much longer than we, the authors, had anticipated. It pushed us to explore questions and ideas that initially we did not realize would be part of the book and to say much more about the relation between a contractarian view of (the ethics of) *war* and a more general contractarian view of morality.

We believe that contractarianism offers an excellent way of thinking about the rules that govern war on both the *ad bellum* and on the *in bello* levels. It sheds light on 'old wars'—those fought between state armies; and on 'new wars'—those fought between states and non-state actors. It introduces unity and logic into the varied rules and principles concerning the law and the morality of war. Whether these beliefs about contractarianism are true or not is now for the reader to judge.

*

Parts of this work have been presented in various forums: The Stockholm Center for War and Peace, The Oxford Institute for Ethics, Law and Armed Conflict (ELAC), Bar-Ilan University, Tel-Aviv University, the Institute for Advanced Study of the Hebrew University of Jerusalem, the University of Haifa, New York University, Georgetown University, Sheffield University, the Schell Center for International Human Rights at Yale Law School, the Institute for Advanced Study at Princeton, the Institute for Advanced Study at Jerusalem, the University of Toronto, the Carnegie Council for Ethics and International Affairs, the Yale Middle East Legal Studies Seminar, the University of Canterbury, the University of Michigan, and Harvard Law School. We are grateful to these institutes for granting us the opportunity to present our ideas and we thank the organizers and participants for their comments.

Over the years, we have accumulated sizable debts to colleagues and students who have discussed contractarianism with us and in some cases were kind enough to send us written feedback. Thanks are due to Dapo Akande, Daniel Attas, Susanne Burri, Gaby Blum, Lars Christie, Roger Crisp, Jeremy Davis, Hanoch Dagan, Avihay Dorfman, Azar Gat, Alon Harel, Moshe Halbertal, Judy Lichtenberg, Ofer Malcai, Larry May, Ariel Porat, Iddo Porat, Massimo Renzo, Avi Sagi, Henry Shue, Daniel Schwartz, Shlomi Segal, Re'em Segev, Scott Shapiro, Saul Smilansky, Victor Tadros, Alec Walen, Joseph Weiler, Jennifer Welsh, Jim Whitman, Elad Uzan, Dror Yinon, and Noam Zohar.

We are especially grateful to Eyal Benvenisti, Cécile Fabre, Owen Fiss, Helen Frowe, Paul Kahn, David Luban, Avishai Margalit, Arthur Ripstein, David Rodin, and Cheyney Ryan, who offered support, encouragement, and penetrating critical feedback on more than one occasion, each in her or his own way.

For the last two decades, Jeff McMahan's influence on the philosophical debate concerning the ethics of war has been decisive. It would be only a minor exaggeration to say that he has transformed the field. Jeff's influence was a result, first and foremost, of his excellent arguments and the clarity of his writing. But his moral passion, friendliness, and generosity no doubt also played a role. No one could hope for a better philosophical opponent than Jeff. We owe special thanks to another revisionist with whom we have had many discussions along the years, Seth Lazar. We are deeply grateful to them both for their readiness to engage seriously in discussions on contractarianism, despite their opposition to its tenets.

Our greatest debt is to Michael Walzer, whose 1977 book *Just and Unjust Wars* is, we believe, still the most important book in the field. We were both privileged to spend time with Michael at the Institute for Advanced Study in Princeton and to discuss with him our thoughts on the ethics of war and on other topics. The combination of deep philosophical thought, rich historical knowledge, and acute understanding of the social and political reality that feed into his writing has always been a model for us, as have his kindness, moral commitment, and modesty. To a large extent, this book can be seen as an interpretation of *Just and Unjust Wars*. We would be very happy if Michael were to accept this interpretation.

In preparing this manuscript, Yitzhak benefitted from sabbaticals at the Schell Center for International Human Rights at Yale Law School and at the Hebrew University of Jerusalem's Institute for Advanced Study. His research was supported by two grants from the Israel Science Foundation (304/15 and 676/09). Danny benefitted from his stay at the Forschungskolleg Humanwissenschaften in Bad Homburg. We both benefitted from the excellent environment at the Shalom Hartman Institute in Jerusalem. We would like to thank our colleagues at the Institute as well as its president, Donniel Hartman, for their support and encouragement.

Two anonymous referees for Oxford University Press gave us excellent comments which helped us improve the manuscript. If they read the final version, they will no doubt identify their traces. It is a bit odd to express gratitude to people without knowing who they are, but we wish to do so nonetheless.

Last but not least, thanks to Anat Shapira for her research assistance, and for Hephzibah Levin for her help in editing the final version of the manuscript. Special thanks to Dr Stella Statman, Danny's mother, for her help both with the language and the content throughout the writing process.

Yitzhak Benbaji and Daniel Statman

Jerusalem,
August 2018

Introduction
The Moral Standing of the War Agreement

The legal prohibition on aggression was first posited in the 1928 Kellogg-Briand pact ('The Pact of Paris'), which outlawed 'war as an instrument of national policy'. The parties to this pact undertook the duty not to use force to resolve 'disputes or conflicts of whatever nature or of whatever origin they may be, which may arise among them'. Later, the United Nations Charter gave expression to the same idea: 'All members shall refrain in their international relations from the threat or use of force against the territorial integrity or political independence of any state.'

The Geneva Conventions explicate the other pillar of the modern law of war. They lay down the rules of warfare: the legal right conferred on combatants to participate in war without vindicating that it is just; the fundamental principles of noncombatant immunity, and the requirement that any side-effects of military actions be subject to proportionality keyed exclusively to military advantage. Famously, these rules symmetrically apply to both sides in a conflict.

As a whole, the legal system thus set out is perplexing mainly because it entitles combatants fighting an unjust war to kill and maim enemy combatants. The puzzlement was well expressed by the British prosecutor at the International Military Tribunal at Nuremberg who noted that:

> [T]he killing of combatants in war is justifiable, both in international and in national law, only where the war is legal. But where the war is illegal, as a war started not only in breach of the Pact of Paris but also without any sort of declaration clearly is, there is nothing to justify the killing; and these murders are not to be distinguished from those of any other lawless robber bands.[1]

---

[1] Quoted in Michael Walzer, *Just and Unjust War* (New York: Basic Books, 2006), 34, and in Arthur Ripstein, *Rules for Wrongdoers: Law, Morality, War* (New York: Oxford University Press, 2021), pp. 20–21. Compare this statement to Francisco de Vitoria's. A political leader, Vitoria says, 'cannot have greater authority over foreigners than he has over his own subjects; but he may not draw the sword against his own subjects unless they have done some wrong; therefore he cannot do so against foreigners except in the same circumstances... It follows from this that we may not use the sword [that is, resort to war] against those who have not harmed us; to kill the innocent is prohibited by natural law' Francisco de Vitoria, 'On the Law of War', in Anthony Pagden and Jeremy Lawrence (eds.), *Political Writings* (Cambridge: Cambridge University Press, 1991), 303–4, quoted in Jeff McMahan 'Just Cause for War', *Ethics and International Affairs* 19 (2005), 4.

The compelling moral idea that underlies this argument is 'that it is individual persons, not states, who kill and are killed in war, and that they, rather than their state, bear primary responsibility for their participation and action in war'.[2] This moral conviction motivates what has come to be known as the 'revisionist' critique of the traditional just war theory.

Thus, according to critics, the legal equality of combatants instituted in the Law of Armed Conflict (LOAC) is reminiscent of the 'regular war view' embedded in pre-1928 international law. According to the regular war view, the society of states is in the condition of war of all against all, whereby each state has the liberty-right to do whatever advances its own interest. The 'procedure' by which disputes are resolved is through a clash between armies.[3] In sharp contrast, the post-1928 law adopted the 'just war view';[4] this is considered to be one of the great achievements in the history of international law. From the just war perspective, a just war is designed to prevent a wrong or to undo it. As such, it is akin to an enforcement action, in which compensation is exacted from, or punishment inflicted on, a wrongdoer.

Now, on the face of it, the just war view implies that Unjust Combatants, fighting against the enforcement of these rights, are wrongdoers—like justly convicted criminals, who use force to avoid proportionate punishment. The legal equality of combatants seems to have no place in the post-1928 international law.

This book addresses the puzzlement caused by this claim. According to the view we defend here, the society of states should organize itself in light of a basic aim that its members ought to pursue together; viz., the preservation of peace. The legal rules regulating resort to and conduct of war express an agreement between the parties on the international level to achieve this aim.

*

In this Introduction we sketch the argument which we present in much greater detail in the course of the book. We hope that this overview will make the relation between the different aspects of the complex argument elaborated in subsequent chapters more transparent. We present it in two stages. We begin with an interpretation of the just war theory embedded in the UN Charter and the LOAC. We suggest that this legal system is best understood as a contract between decent states whose aim is to maintain the peace of the status quo ante. As part of this contract, states agree to outlaw the use of force. They waive their pre-contractual moral right to use force even where force is necessary to achieve certain just aims. And they allow each other to go to war in defence of their

---

[2] Jeff McMahan, 'Rethinking the "Just War"', Part 1, *New York Times Opinionater*, November 11, 2012. (https://opinionator.blogs.nytimes.com/2012/11/11/rethinking-the-just-war-part-1/).

[3] We borrow this exposition from Arthur Ripstein, *Kant and the Law of War* (forthcoming), Chapter 1. The regular war view is attributed to Hugo Grotius and his followers.

[4] Ripstein attributes this view to Thomas Aquinas and the Salamanca scholastics, Francisco de Vitoria and Francisco Suárez.

contractual right against the first use of force. In the second stage, we argue that legal systems in general, and the war agreement in particular, are morally binding (or 'morally effective') if they pass certain moral tests. By accepting the rules that constitute these legal regimes, states, as well as their citizens, lose some of their pre-contractual moral rights and gain other contractual moral rights.

Before continuing, we should note that we are not the first to take the contractarian route in seeking to understand and justify the ethics of war.[5] George Mavrodes published the pioneering article in this regard some forty years ago.[6] Here and there, Walzer also uses contractarian arguments, arguing that 'military conduct is governed by rules [that] rest on mutuality and consent'.[7] More recently, we find such arguments in writers such as Jeremy Waldron[8] and Tom Hurka.[9] Surprisingly, even Jeff McMahan, the leading critic of the traditional theory, shifts sometimes to a contractarian mode of reasoning, suggesting that it is rational for each side in a conflict to adhere to the conventions of war if the other side does. The adherence must be mutual. 'Thus if one side breaches the understanding that the conventions will be followed, it may cease to be rational or morally required for the other side to persist in its adherence to them.'[10]

Yet, none of these writers has consistently followed contractarian logic as a basis for the rules regulating war, which is what we endeavour to do here.[11] If we succeed, then finally, after two decades of perplexity regarding the philosophical foundations of the accepted war convention, its moral logic can be appreciated.

The War Agreement: *Jus ad Bellum*

According to contractarianism, the UN Charter (in which the traditional conception of *jus ad bellum* is embedded) and the LOAC (in which the traditional conception of the *jus in bello* is embedded) are best understood as two aspects of an agreement between decent yet partial states.

---

[5] It has become common to distinguish between 'contractarianism', which refers to Hobbes, Gauthier, and their followers, and 'contractualism', which refers mainly to Scanlon. See Elizabeth Ashford and Tim Mulgan, 'Contractualism', *Stanford Encyclopedia of Philosophy*, 2012. We identify more with the former camp.

[6] George I. Mavrodes, 'Conventions and the Morality of War', *Philosophy and Public Affairs* 4 (1975), 117–31.

[7] Walzer, *Just and Unjust Wars*, 37.

[8] Jeremy Waldron, *Torture, Terror and Trade-Offs: Philosophy for the White House* (New York: Oxford University Press, 2010), 80–110.

[9] Thomas Hurka, 'Liability and Just Cause', *Ethics and International Affairs* 20 (2007), 199–218. See also George P. Fletcher and Jens D. Ohlin, *Defending Humanity: When Force is Justified and Why* (New York: Oxford University Press, 2008), 180, claiming that the principle of reciprocity 'governs the entire law of armed conflict'.

[10] Jeff McMahan, 'The Ethics of Killing in War', *Ethics* 114 (2004), 730.

[11] The strong reliance of the rules governing war on the notions of mutuality and reciprocity might explain the (by and large) higher level of compliance with these rules than with human rights laws. See Eric Posner, 'Human Rights, the Laws of War, and Reciprocity', *Law and Ethics of Human Rights* 6/2 (2012), 147–71.

In forming the *ad bellum* agreement, states aim to minimize the use of force—however justified the use of force may be—in the international society. As Article 1 of the UN Charter asserts, the United Nations aims '[t]o maintain international peace and security' by subjecting states to the duty to resort to peaceful means in resolving 'international disputes or situations that might lead to a breach of the peace'. In jointly pursuing this end, members of the United Nations undertake the duty set out in Article 2(4) of the Charter, not to use force against the territorial integrity and sovereignty of another state.

Although, in rare cases, wars aimed at attaining a fairer distribution of vital resources (viz. just subsistence wars) are pre-contractually permissible, by subjecting themselves to the Article 2(4) prohibition, decent states waive their moral right to initiate such wars. Similarly, in rare cases, waging a preventive war in order to eliminate an immature threat is pre-contractually permissible. Yet, in adhering to the *ad bellum* agreement embedded in the Charter, states waive their right to wage such wars.

As with any system of legal rules, the war agreement includes rules about non-compliance.[12] The 2(4) prohibition on the use of force is enforced by self-help, through the Article 51 right to fight defensive wars. States allow each other to engage in war against those who violate their contractual right to territorial integrity. Once they engage in such defensive wars, the defenders are allowed to continue fighting in order to deter the actual aggressor (and, as a side-effect, potential aggressors) from aggression against it in the future, by making aggression costly to the aggressor. This permission, however, is not unlimited. Attaining total certainty that the enemy will not be able to use force in the future would require the waging of a ruthless war, which would be too costly to both sides.

In sum, according to our reading of the *ad bellum* agreement, the Charter outlaws any use of force whose aim is a resolution of a political dispute, even if the state contemplating it has a justified grievance against its adversary and the capacity to enforce its rights by force. At the same time, the Charter permits states to use force in defence of their borders without committing itself to the justice of the status quo ante. Aggression—the crime against peace—is a disruption of the status quo ante by means of war. But the peace that the Charter aims to maintain is just a peace, not necessarily a just peace.[13]

The War Agreement: *Jus in Bello*

We turn now to the *in bello* contract. As we suggest reading it, the goal of this part of the war agreement is to enable states to effectively defend their rights while minimizing the horrors and the casualties of war. More specifically, the *in bello*

---

[12] Brad Hooker, *Ideal Code, Real World: Rule-Consequentialist Theory of Morality* (Oxford: Oxford University Press, 2000), 82.
[13] Avishai Margalit, *On Compromise and Rotten Compromises* (Princeton, NJ: Princeton University Press, 2010), 1.

agreement entitles the parties to the contract to develop the means to maintain peace. As such it allows states to hold military forces that will deter potential aggressors from violating the contractual right to their independence and territorial integrity, as well as to wage defensive wars against actual aggressors who have already violated these rights.

Relying on some empirical assumptions regarding the nature of states and of the society in which states are 'united', the *in bello* contract entails the legal equality of soldiers and of civilians. The legal equality of soldiers implies that soldiers are allowed to follow an order from their political leaders to participate in war without ensuring that the war is just or legal. What underlies this permission is the assumption that requiring soldiers to make sure that their war is just would weaken the ability of states to deter their enemies from aggression and to effectively protect their rights in case aggression erupts. States, therefore, grant each other the right to maintain obedient armies, whose soldiers are responsible only for their own conduct in war, not for the war itself. According to our reading of the LOAC, soldiers on both sides lose their legal right not to be attacked by each other in exchange, so to speak, for a legal right to attack each other without first making sure that their war is just.

Let us turn to civilian immunity. If the agreement is accepted, decent states attack one another and defend themselves against such aggression by using the military forces at their disposal. If the agreement is accepted, targeting civilians does not typically promote the just aims of war, states undertake a general duty not to target civilians. True, in some cases, targeting civilians might be an effective way to eliminate an unjust threat—particularly in the case of civilians who are responsible for the threat posed by their country. Still, posing a blanket prohibition on targeting civilians would be *ex ante* better for all parties. It would minimize the casualties of war without compromising the capability of defenders to block what they regard as unjust aggression against them.

Our interpretation of the war agreement addresses most aspects of war—its initiation (*jus ad bellum*), its aims (which implies a theory about the termination of war, *jus ex bello*), and its conduct (*jus in bello*). We seek to show that understanding each of these aspects as an agreement between states sheds light on the current arrangements. However, we disregard here one aspect of war that has received a lot of attention in the last decade; namely, justice-after-war (*jus post bellum*).[14] Since the *jus ad bellum* and the *jus in bello* issues comprise the 'main event' in the debate between traditionalists and revisionists, they will be the focus of our present attention. Extending the contractarian framework to issues of *post bellum* is a worthy enterprise for another occasion.

---

[14] See e.g. Brian Orend, 'Justice After War', *Ethics and International Affairs* 16/1 (2002), 43–56, and Carsten Stahn, Jennifer S. Easterday, and Jens Iverson (eds.), *Jus Post Bellum: Mapping the Normative Foundations* (Oxford: Oxford University Press, 2014).

The Moral Standing of a Sufficiently Good Legal System

Our interpretation of the war agreement is based on a set of empirical assumptions that the revisionists tend to reject. In particular, revisionists believe that a regime that compels soldiers to make sure that their war is just as a condition for their participation in it better protects peace and justice than the current regime, which allows them to participate in war without posing such a demand. Similarly, some revisionists believe that a regime that allows the intentional killing of civilians who are culpable for the wrong against which the just side is fighting better protects peace and justice than does a regime that treats all civilians as immune to direct attack.

Thus, the debate between revisionism and contractarianism is partly empirical. Traditional just war theory—as contractarianism restructures it—sees wisdom in realist pessimism regarding the insecure environment in which national communities exist. Realism in international relations is first and foremost a descriptive theory about the chronic insecurity in a global society that lacks any central authority to prevent crimes against peace. Contractarianism fully embraces this factual observation. Accordingly, we assume that strong armies—viz., armies whose soldiers are obedient and as such easier to activate—make the world more secure rather than less so. (Compare: obedient police forces make domestic society more secure, even if the police force abuses its rights to use force from time to time.) Most revisionists deny this assumption.

Yet, the most fundamental conviction that underlies the revisionist view is normative rather than empirical. Revisionists insist that the justice of wars and the morality of killing in them are independent of the laws of war. Wars cannot be (overall) morally justified merely because an optimal legal system allows states to fight them and they cannot be immoral merely because a good or optimal legal system prohibits them. Similarly for the *in bello* level. Killing 'Just Combatants'— those soldiers who fight with a just cause—cannot become morally permissible even if an optimal legal system allows such killings. Similarly, killing culpable civilians cannot become morally impermissible merely because the optimal legal system prohibits it.

By contrast, contractarianism presumes that optimal legal systems, or even sufficiently good actual ones, are morally effective. When a legal system is good enough, the consent of the parties to be governed by it is given freely—in which case it involves a waiver of the relevant pre-contractual rights. This contractarian assumption has far-reaching implications. At the *ad bellum* level, it implies that non-defensive wars are immoral (or unjust) since states undertake the duty not to fight them and that defensive wars are permissible (or just) since states allow each other to fight them. At the *in bello* level it implies (among other things) that killing soldiers in war does not wrong them whatever the cause of the war is, since both Just and Unjust Combatants subject themselves to a regime that equalizes their legal standing. Similarly, killing civilians is impermissible, since soldiers and their leaders subject themselves to a legal system that grants immunity to civilians.

Thus, contractarianism is based on the assumption that states, soldiers, and civilians have the moral power to waive their pre-contractual rights by subjecting themselves to a sufficiently good legal system. This assumption stands at the heart of the morality of the relations between citizens and their own states as well as between states and other states.

Many moral philosophers believe that the role of moral theory is to spell out the right- and wrong-making features of actions ('conducive to overall utility', 'respectful of human rights', and so on). They assume that when these features are properly identified, one can know whether actions are morally acceptable or not. You may call this approach 'act-focused morality'; morality provides tools to directly ascertain the status of each act and determine whether it is permissible, mandatory, or forbidden. In contrast, according to contractarianism, in many social circumstances, morality should focus on the rules to be followed rather than directly on the morality of the acts—which is judged secondarily, according to whether the act complies with the rules. The fact that moral thinking about war in the last two decades has been dominated by act-focused morality is a major reason for the stalemate mentioned earlier in the philosophical discussion over war. We believe that the rule-based perspective of the sort we develop in this book is an important step toward overcoming it.

The Structure of the Book

A few last words on the structure of the book. In Chapter 1, we lead the reader through a quick tour of the current debate on the ethics of war. We explain the views of those known as traditionalists and those known as revisionists. We present the criticism mounted by the revisionists against Walzer's interpretation of the legal system that regulates warfare, and then show that the alternative proposed by revisionists is problematic in a number of respects.

Chapter 2 lays down the fundamental normative assumption that distinguishes contractarianism as we interpret it from revisionist individualism. We analyse a series of cases that show that some of the moral duties to which individuals are subject and some of the moral rights that they bear are grounded in social rules that they freely accept. If a set of social rules is fair and mutually beneficial, then by actually accepting it, these individuals waive the rights that they would enjoy vis-à-vis others in the state of nature. These cases show that the distribution of moral rights and duties in actual societies may differ radically from their pre-contractual distribution.

The rest of the book is an application of this theoretical framework to the context of war. Chapter 3 addresses the *ad bellum* agreement. It interprets the rules governing the initiation of wars as part of an agreement between states in what we refer to as 'minimally just symmetrical anarchy'. It is in the interest of states to waive their right to use force in resolving their disputes, even if on rare occasions they are morally justified in using force to this aim. According to our reading of the Charter's 'inherent right to self-defense' clause, states agree to allow defensive wars against any illegal use of force, no matter how justified the illegal war might have been without the agreement. Finally, we observe that the logic of the *ad bellum* agreement implies that in some cases, stateless nations that require political independence have a just cause for a war of independence.

Chapter 4 deals with the aims of the just war and, by implication, with what has come to be known as *jus ex bello*. We argue that although defensive wars are basically a remedy, they might be fought until the defender is certain that the aggressor has no potential to resort to another aggressive war in the near future. The permission has an important side-effect: it helps to deter potential aggressors by allowing the defender to impose extra costs on them. Accordingly, the Charter confers on states a derivative right to maintain strong armies that will enable them to impose these costs.

Chapters 5 and 6 focus on *jus in bello*, the rules concerning conduct of war. We show that the contractarian logic explains well the basic principles in this field; the distinction between combatants and noncombatants, the blanket permission to attack enemy combatants, the blanket prohibition on targeting civilians, and the permission to cause collateral harm to civilians. These chapters appeal to another normative assumption that contractarianism centralizes; namely, individuals freely consent to be governed by legal rules accepted in the society to which they belong, as long as these rules were posited by a state that acts on their behalf. This understanding applies to the domestic law by which their society is governed, as well as to the treaty-based international law on which states agree.

Chapter 7 turns to the question of what happens when the war agreement collapses. This question bothers many people in the free world today, as we face barbaric attacks by rogue organizations such as Al-Qaeda and ISIS. This chapter offers three claims. The first addresses the *ad bellum* level. As far as possible, just states ought to treat unjust (aggressive) states or non-state actors as if these aggressors mistakenly believe themselves to be fighting for a just cause, such as national independence, political freedom or de-colonization. The second claim addresses violations of the *in bello* agreement. Such violations do not return the parties to the state of nature in which their war would be governed by pre-contractual morality. Rather, the war agreement should be read as dictating responses to these violations by conferring remedial rights whose scope is strictly restricted. Third, in supreme emergencies, where the aggressor is clearly indecent and poses an existential threat that cannot be addressed by conventional war, the parties are entitled to use weapons of mass destruction in addressing the threat. Chapter 8 then offers some concluding remarks.

1
The Challenge

1.1 Introduction

The UN Charter and the Geneva Convention have shaped the way that most people today think about 'right' and 'wrong' in war; not only in the narrow legal sense, but in a moral sense as well. In this field, the legal and ethical domains seem to overlap to a large extent.[1] This close relation between the ethics of war and the laws of war received classic expression in Michael Walzer's 1977 *Just and Unjust Wars*, which for decades has been the bible, so to say, of the ethics of war. In Walzer's view, the laws of war are 'morally plausible',[2] which means that following them is usually sufficient to being morally right, both for the relevant individuals (mainly combatants) and for states.

This view, referred to hereafter as the 'traditional view',[3] came under serious attack a decade or so ago from a group of philosophers led by Jeff McMahan, who have become known as 'revisionists'.[4] According to these philosophers, the traditional view has failed to substantiate the moral underpinnings of the accepted legal arrangements for war; or, put in other words, it has failed to make the

---

[1] Although in some contexts a distinction does exist between 'moral' and 'ethical', in the present context it does not, so we use these terms interchangeably.

[2] Walzer, *Just and Unjust Wars*, 133.

[3] In addition to Walzer, Seth Lazar mentions the following thinkers as belonging (fully or partially) to this group: Margaret Moore, 'Collective Self-Determination, Institutions of Justice, and Wars of National Defence', in Cécile Fabre and Seth Lazar (eds.), *The Morality of Defensive War* (Oxford: Oxford University Press, 2014), 185–202; Yitzhak Benbaji, 'A Defense of the Traditional War Convention', *Ethics*, 118/3 (2008), 464–95; Noam Zohar, 'Collective War and Individualistic Ethics: Against the Conscription of "Self-Defense"', *Political Theory* 21/4 (1993), 606–22; Christopher Kutz, 'The Difference Uniforms Make: Collective Violence in Criminal Law and War', *Philosophy and Public Affairs*, 33/2 (2005), 148–80; and Janina Dill and Henry Shue, 'Limiting the Killing in War: Military Necessity and the St Petersburg Assumption', *Ethics and International Affairs* 26/3 (2012), 311–33. See Seth Lazar, 'War', *Stanford Encyclopedia of Philosophy*, 2016.

[4] Prominent members of this group include Jeff McMahan, 'The Ethics of Killing in War', *Ethics* 114 (2004), 693–733; David Rodin, *War and Self-defense* (New York: Oxford University Press, 2002); Cécile Fabre, *Cosmopolitan War* (Oxford: Oxford University Press, 2012); Helen Frowe, *Defensive Killing* (Oxford: Oxford University Press, 2014); Adil A. Haque, 'Law and Morality at War', *Criminal Law and Philosophy* 8 (2014), 79–97; Charles R. Beitz, 'Nonintervention and Communal Integrity', *Philosophy and Public Affairs* 9 (1980), 385–91; Richard Norman, *Ethics, Killing and War* (Cambridge: Cambridge University Press, 1995); Lionel McPherson, 'Innocence and Responsibility in War', *Canadian Journal of Philosophy* 34/4 (2004), 485–506; Richard J. Arneson, 'Just Warfare Theory and Noncombatant Immunity', *Cornell International Law Journal* 39 (2006), 663–88; See Lazar, 'War', *Stanford Encyclopedia of Philosophy*, 2016.

required distinction between the legal and the ethical realms in issues concerning war. Revisionists thus propose a fresh look at the moral principles that govern the behaviour of states and individuals in war.

One prominent revisionist approach bases the morality of war on the principles of individualist morality. This approach is referred to as 'reductionist' because it seeks to reduce all moral normative truths about war to normative truths about individuals. According to the revisionists, the resort to war and its conduct must be justified by appeal to the same principles that govern the relations between individuals in non-war contexts.[5]

The revisionist arguments have revitalized the field and led to a flood of books and articles on the ethics of war, comprising an occurrence not far from a paradigm shift. While in the 1980s and 1990s Walzer's view generally defined the paradigm within which debates about the ethics of war were conducted—designated as the 'orthodoxy', so to speak—now the revisionist view has become paradigmatic in conferences and in the literature; in fact, it has become the new orthodoxy.

This book is motivated by the belief that while the revisionist criticism of the traditional view is convincing, the alternative it proposes is less so. The goal of the present chapter is to elaborate on these two points; namely, to present the revisionist critique of the traditional view of war—viz., Walzer's moral reading of the accepted regulations on the use of force—and then show how revisionism fails to provide an alternative to traditional just war theory.

In Section 1.2 we present briefly the main ideas of the traditional view. In Section 1.3 we introduce *Individualism*, which we take to be the prominent moral outlook assumed by most revisionists. In Sections 1.4 and 1.5 we present the critique offered by *Individualism* on both the *ad bellum* and the *in bello* levels and show why the alternative it offers fails.

1.2 The Traditional View

The traditional understanding of the ethics of war starts with a distinction between two separate judgements about war: judgements regarding the launching

---

[5] This wave is not completely new. Soon after the publication of *Just and Unjust Wars*, Walzer addressed a similar critique against the prohibitive *jus ad bellum* of the UN Charter. His four critics argued that instead of defending individuals, the Charter (as well as Walzer's moral reading of it) defends both states and the communities they represent. See Richard Wasserstrom, 'Review of Michael Walzer's *Just and Unjust Wars: A Moral Argument with Historical Illustrations*', *Harvard Law Review* 92/2 (1978), 536-45; Gerald Doppelt, 'Walzer's Theory of Morality in International Relations', *Philosophy and Public Affairs* 8/1 (1978), 3-26; Charles R. Beitz, 'Bounded Morality: Justice and the State in World Politics, *International Organization* 33/3 (1979), 405-24; David Luban, 'Just War and Human Rights', *Philosophy and Public Affairs* 9/2 (1980), 160-81; and Walzer's response at Michael Walzer, 'The Moral Standing of States: A Response to Four Critics', *Philosophy and Public Affairs* 9/3 (1980), 209-29.

of war and regarding the way war is conducted. The first concerns the justice *of* war, *jus ad bellum*, and the latter justice *in* war, *jus in bello*. These judgements, says Walzer, 'are logically independent. It is perfectly possible for a just war to be fought unjustly and for an unjust war to be fought in strict accordance with the rules.'[6] This proposition has come to be known as the independence thesis, hereafter *Independence*. Its main implication is that combatants of the unjust side ('Unjust Combatants', those on the side unjustified in going to war) have a right to shoot combatants of the just side ('Just Combatants', those on the side that is justified in doing so). As long as combatants fight in accordance with the rules of war, they act within their rights, even if the war in which they participate is unjust.

Let us start with the principles of *jus ad bellum*. Article 2(4) of the UN Charter dictates that 'all members shall refrain in their international relations from the threat or use of force against the territorial integrity or political independence of any state'. If this article is violated and some state is subject to such a threat, it needs not wait until the UN sends troops to help, but is allowed to use force to block the attack. Members of the United Nations—namely, states that are recognized by other states as legitimate—have an 'inherent right of individual or collective self-defence if an armed attack occurs against' them. Article 51 should be read against Article 2(4) and interpreted as referring to the use of force directed 'against the territorial integrity or political independence of any state'.

Threats to territorial integrity or political independence thus comprise a necessary condition for the legality of wars, not a sufficient one. Other conditions must be satisfied, most importantly necessity and proportionality but also probability of success, legitimate authority, and rightful intention. War must comprise the 'last resort' in response to such threats, after diplomatic and other non-violent measures have been exhausted. Moreover, the expected gain from the war must be proportionate to the anticipated death and harm that will be caused. Thus, for example, the potential loss of some unoccupied tiny island with no strategic importance is probably not worth a full-scale war.[7]

From a moral perspective, why are states allowed to go to war in the face of such threats? Arguably, because their most fundamental interests are in danger. First, the fundamental interest of their citizens in their own lives. As Walzer puts it, 'once the lines are crossed, safety is gone'.[8] Thus, to protect the safety of their citizens, states are allowed to take up arms and fight against their enemy. Second, their interest in the political community that is shaped by the state and its citizens.

---

[6] Walzer, *Just and Unjust Wars*, 21.

[7] Quoting from a diplomatic note exchanged between US Secretary of State Daniel Webster and British Privy Counselor Alexander Baring, Lord Ashburton, in 1841, Michael Doyle explains that an attack must: (1) be 'overwhelming' in its necessity; (2) leave 'no choice of means'; (3) come in response to such an imminent threat that there is 'no moment for deliberation'; and (4) be proportional (Michael W. Doyle, *Striking First: Preemption and Prevention in International Conflict* [Princeton, NJ: Princeton University Press, 2008], 12).

[8] Walzer, *Just and Unjust Wars*, 57.

According to Walzer, an organized political community is 'conceivably the most important good', as it enables people to 'share a way of life, developed by their ancestors, to be passed on to their children'.[9]

Let us turn now to the three main principles constituting *jus in bello*:

(a) With the exception of POWs, and other particular groups, combatants in war have an almost unlimited right to attack enemy combatants intentionally.
(b) Combatants may not intentionally attack noncombatants, even if the expected military advantage of such an attack is substantial and the costs in terms of harm to civilians slight (hereafter: *Civilian Immunity*).
(c) Combatants may bring about harm to noncombatants as a side effect of attacks on military targets, provided that the harm is necessary and proportionate to the military goal, and that an effort is made to minimize civilian casualties (hereafter: *Collateral Damage*).

These principles of the traditional *jus in bello* code grant moral equality to combatants on both sides and to noncombatants on both sides.[10] Regardless of whether they fight for the just or the unjust side, combatants have a right to attack combatants of the other side; and regardless of whether their state is fighting for a just or an unjust cause, noncombatants are morally and legally immune from deliberate attacks against them. Both sides have no right against incidental, proportionate harm to civilians. We refer to this principle as *Moral Equality*.[11]

Moral Equality does not entail that combatants of the unjust side are morally equal in all aspects to those fighting on the just side. After all, the former are fighting for an unjust cause while the latter are merely defending themselves from an unjust attack. Similarly with noncombatants; surely they are not equal in all morally relevant respects. Civilians belonging to the unjust side ('Unjust Civilians') are often implicated in a variety of ways in an unjust project, while those on the just side are not. *Moral Equality* is mainly about duties, claim-rights, privileges and immunities. Combatants have a right to fire at their adversaries—they violate no duty toward their victims by doing so, whether the war they fight is just or not. Walzer insists that the legality of the intentional killing of combatants and the legality of incidental killing of civilians reflect the morality of killing in war;

---

[9] Walzer, 'The Moral Standing of States: A Response to Four Critics', 212.

[10] As Cheyney Ryan reminds us, just war theorists did not always adhere to this perspective. Those writing in the Middle Ages, for instance, did not regard war as a conflict between equals. On the contrary, killing in war was seen as just punishment for the sinful aggression of the other side. See Cheyney Ryan, 'Democratic Duty and the Moral Dilemmas of Soldiers', *Ethics* 122 (2011), 14–15, referring to Stephen Neff, *War and the Law of Nations* (Cambridge: Cambridge University Press, 2005).

[11] The notion is expressed in the Geneva Convention, which states that the provisions of the convention apply to all persons 'without any adverse distinction based on the nature or origin of the armed conflict or on the causes espoused or attributed to the parties of the conflict'; see Protocol Additional to the Geneva Conventions of 1949, and relating to the Protection of Victims of International Armed Conflicts 1977, 1125 UNTS 3.

combatants seeking to promote a legitimate military goal have the moral right to deliberately attack combatants and to harm civilians incidentally (if such harm is necessary and proportionate). Combatants who comply with the rules of war stated above do not violate any duty they bear towards others. The same is true of noncombatants. Even if they belong to the unjust side ('Unjust Noncombatants'), they benefit from the right not to be deliberately attacked; in this sense, they are no different from their counterparts on the just side ('Just Noncombatants').

This *Moral Equality* is sometimes referred to as the 'symmetry thesis', stating that the content of the *jus in bello* rights and obligations is the same for combatants on both sides of any conflict. Conversely, the 'asymmetry thesis' states that the content of *jus in bello* rights and obligations is not the same for combatants on both sides.[12] *Moral Equality* is the flip side of *Independence*; if the rights and obligations that apply to combatants are independent of the justness of the cause for which they are fighting, then combatants of all sides are subject to the same rights and obligations. Conversely, if moral equality exists between combatants of all sides, that entails the independence of judgements on the *in bello* level from judgements on the *ad bellum* level.[13]

The logical relation between *Moral Equality* and *Independence* leaves its directionality open; one could start with *Moral Equality* and deduce *Independence* from it, or vice versa. As we clarify below, we hold the latter view; namely, that solid grounds support *Independence*, which, in turn, justifies *Moral Equality*.

Revisionists challenge all the above statements about the morality of warfare. Before we explain this challenge, we present in the next section what we take to be their fundamental moral commitment. This understanding will help elucidate the revisionist 'heretical' views on the ethics of war.

1.3 Individualism and Continuity

The critique of the traditional view relies on a moral outlook that we dubbed *Individualism*. In this section, we introduce this outlook.

The traditional view takes the ethics of war to be reflected in the *laws* of war. Most individualists strongly reject this close relation between the laws of war and the morality of war. In their view, even if the current legal regime that governs war is optimal and accepted by all parties, it falls short of defining the *moral* duties and rights of individuals during times of war. In particular, the fact that combatants violate no legal rule when they kill enemy combatants even when in fighting for an

---

[12] Walzer, *Just and Unjust Wars*, 36.
[13] David Rodin and Henry Shue, 'Introduction', in David Rodin and Henry Shue (eds.), *Just and Unjust Warriors: The Moral and Legal Status of Warriors* (Oxford: Oxford University Press, 2008), 3.

unjust cause does not mean that they do not violate any moral prohibition.[14] What seems to underlie this objection is a general view about the centrality of individual moral rights, which has implications for all branches of applied ethics. Here is a first attempt to characterize this notion:

Individualism: The moral duty incumbent on each person to respect the most fundamental human rights of all other persons does not depend on the national, religious, or other affiliation of the person or of the right-bearer. Nor does it depend on the social role that any person might happen to have, qua citizen of a specific state, combatant of a particular army, or bearer of a specific role (such as policewoman, mayor, judge, banker, etc.).

In using the term *Individualism*, we try to capture the assumed stability of fundamental individual rights that, from this perspective, are not subject to exchange within social arrangements. Seen this way, the fundamental moral relations that exist between human beings are often obscured by the commitments and loyalties that people have by virtue of their contingent affiliations and roles.[15]

Note that *Individualism* is not committed to the view that human rights may never be infringed; namely, to *Absolutism*.

Absolutism: Under no circumstances is it permissible to infringe upon fundamental human rights (for instance to torture people, to kill the innocent, and so on).

Very few philosophers (or lay people) subscribe to this extreme view. The more common view is that rights pose significant constraints on the promotion of desirable outcomes (however defined), but that when the stakes are high enough, rights give way. This is true of *Individualism* as well. What this notion proposes is not a different balance between rights and consequences, but a view about the nature of rights; or, more accurately, about the relation between fundamental human rights and commitments or permissions that stem from social context. *Individualism* insists that it is these rights that mark the deep normative structure of human relations, whereas the normative results of our social context are secondary and conditional.[16] It insists that fundamental rights are natural, in the sense that individuals possess these rights independently of any social

---

[14] This separation between the morality of war and the laws of war is central to McMahan's approach. See, for instance, Jeff McMahan, 'The Morality of War and the Law of War', in Rodin and Shue (eds.), *Just and Unjust Warriors*, 35.

[15] In earlier work, we referred to this as 'purism' which was supposed to express this uncompromising commitment to individual human rights. Since this term sounds a bit pejorative, we decided to replace it by '*Individualism*'.

[16] Our use of '*Individualism*' is stipulative and rather technical. As we clarify in Chapter 3, in a non-technical use of the term, the approach we advocate might be described as individualist because of the central role it assigns to individuals' tacit acceptance of social norms.

arrangement. More importantly, it insists that no social arrangement, social rule, or social role can impact the distribution of these rights.

It follows from these commitments that there is nothing morally unique about wars. Just wars are defensive acts that are justified by virtue of the same principles that justify defensive killing and maiming in other contexts. The same conditions that apply to defensive killing in one versus one situations apply also in two versus two situations, or a thousand versus a thousand—or an entire army versus another army. True, the aggressors and victims shooting at each other are *combatants*, instruments of the state on behalf of which they are fighting; but this fact makes no real moral difference. Those commitments lead to *Continuity*.

Continuity: 'The morality of defense in war is continuous with the morality of individual self-defense. Indeed, justified warfare simply is the collective of individual rights of self- and other-defense in a coordinated manner against a common threat.'[17]

According to *Continuity*, when war breaks out we don't 'shift gears', as it were, to a different mode of moral thinking (in contrast to a seemingly common view).[18] We remain subject to the same principles that govern human relations in times of peace; in particular, those governing the right to act in self-defence.

The problem is that when principles of individual self-defence are applied to the context of war, they lead to completely different conclusions from those endorsed by the traditional view. The next sections are devoted to a discussion of these conclusions—first at the level of *jus ad bellum* and then at the level of *jus in bello*. At each level, we start by presenting the individualist critique of the traditional view. We then describe the alternative proposed by *Individualism*, and finally we explain our dissatisfaction with this alternative.

1.4 *Individualism* and the Moral Justification to Wage War

Objections to Traditional *Jus ad Bellum*

We saw above that international law only licenses wars that address threats to territorial integrity or political sovereignty (and, in some cases, wars of national

---

[17] McMahan, 'The Ethics of Killing in War', 717. As Fletcher and Ohlin note, philosophers and lawyers use the term 'collective' in the context of war in different ways: 'What philosophers call "collective self-defence" (i.e. the defence of the nation) is simply called "individual self-defence" by the international lawyers, a term that philosophers reserve for self-defense exercised by individual human beings.' See George P. Fletcher and Jens David Ohlin, Defending Humanity: When Force is Justified and Why (New York: Oxford University Press, 2008), 24.
[18] For anecdotal illustrations, see Andrea Tantaros, 'This is war, not a law seminar: Obama is more Carter than Churchill', *New York Daily News*, 30 September 2010, available at http://www.nydailynews.com/opinion/war-not-law-seminar-obama-carter-churchill-article-1.443668?barcprox=true.

liberation as well). At a first glance, revisionism seems to imply that this licence is both too permissive and too restrictive.

International law is too permissive because, as Richard Norman and David Rodin have shown,[19] 'defensive wars'—wars fought in defence of territorial integrity or political independence—are not necessarily fought in defence of the life of any individual. Fighting in defence of territorial integrity and political liberty by killing and letting others kill seems disproportionate. Just as in individual self-defence, the evil brought about by the defensive act must be proportionate to the evil prevented; in national defence too, the evils brought about by the act of war must be proportionate to the values defended. But as valuable as territorial integrity and political sovereignty are, they are not valuable enough to justify the large-scale killing and the vast destruction that is typically brought about by war. Traditional just war theory, Norman concludes, 'does not succeed in its primary aim. It does not provide a way of rebutting the initial moral presumption against war in any form.'[20] Rodin reaches the same conclusion, asserting 'that the conception of a moral right of national-defence cannot, in the final analysis, be substantiated'.[21]

Thus, although it is almost always legally permissible for states to go to war in order to defend their borders from invasion, *Individualism* seems to imply that from a moral perspective, they usually may not do so. By morally licensing such wars, therefore, the traditional view is over-permissive.

The traditional *ad bellum* approach is also over-restrictive, because of its built-in bias in favour of states. As mentioned above, the UN Charter licenses defensive wars only when *states* are under threat, not when *individuals* are under threat. As international law is usually understood, states rely on the assumption that it takes an army to fight an army, and it confers the right to have armies only on states. Heroic little bands have no right to engage in war, for nearly any cause. International law does not tolerate vigilante intervention presumably because 'it gets hard in principle to distinguish self-appointed saviors...from mafias'.[22]

The problem with this legal system is that individuals might be under threat from their own countries or from some natural catastrophe, such as hunger, which their own or another state is under duty to prevent. Moreover, although the Security Council could take measures to help individuals who face such threats, according to Article 51 of the Charter, such individuals would not have a right to go to war to defend themselves against violations of their most fundamental rights.[23]

---

[19] See Norman, *Ethics, Killing and War*, 120–36, and Rodin, *War and Self-Defense*, 122–38.
[20] Norman, *Ethics, Killing and War*, 206.
[21] Rodin, *War and Self-Defense*, 196.
[22] David Luban, 'Intervention and Civilization: Some Unhappy Lessons of the Kosovo War', in Pablo de Greiff and Ciaran Cronin (eds.), *Global Justice and Transnational Politics* (Cambridge, MA: MIT Press, 2002), 84–5.
[23] See Yitzhak Benbaji, 'Legitimate Authority in War', in Helen Frowe and Seth Lazar (eds.), *Oxford Handbook of Ethics of War* (Oxford: Oxford University Press, online edition, 2015).

1.4 INDIVIDUALISM AND WAGING WAR

David Luban expresses a similar objection to the focus on states embedded in the UN Charter. Consider his example involving Poor and Wealthy, states that share a long border. A part of the border that Wealthy does not share with Poor runs along the ocean; Wealthy, therefore, receives plentiful rainfall. Poor is relatively dry. Mountains prevent rain clouds from crossing over to Poor, whose climate consequently is semi-arid. Several years of drought have caused famine in Poor, threatening the lives of millions. (In an enriched and very realistic example, Poor's grim situation is a result of colonialism or of other injustices it suffered historically or still suffers today.)[24]

Assuming that the deprivation is not Poor's responsibility, if Wealthy can provide Poor with the water needed without undermining its own ability to survive and to flourish, it has a moral duty to do so. Moreover, on the surface at least, it seems that if Wealthy refuses, the people living in Poor have a right to take arms against Wealthy and try to obtain by force the resources to which they are entitled. *Individualism* seems to imply that if they cannot do so themselves, other states (or other individuals!) have a right to wage war against Wealthy on their behalf. Following Fabre, we shall call such wars 'subsistence wars',[25] and regard them as the paradigmatic example of what we shall call 'justice-implementing wars'; wars concerned not with self-defence (narrowly construed) but with promoting global distributive justice.[26]

By ruling out all subsistence wars, the traditional view is thus over-restrictive. While it permits wars that typically do *not* involve a serious violation of fundamental human rights, it forbids wars that sometimes *do* seem to involve such violation. Put differently, the traditional view is too permissive, since it allows wars that aim at the defence of states even if these wars involve killing but do not protect basic human rights. It is too restrictive because it rules out wars that aim at the protection of such rights in a much more direct manner, such as subsistence wars.

Most famously, though, the attitude of the current regime embedded in the Charter and Law of Armed Conflict (LOAC) to wars of humanitarian intervention seems distorted from an individualist point of view.[27] When large-scale violations of human rights take place, *Individualism* seems committed to license the launching of war to block such violations; furthermore, it actually *encourages* the initiation of war in the name of justice. But this individualist impulse is not

---

[24] Luban, 'Just War and Human Rights', 178.
[25] Fabre, *Cosmopolitan War*, chapter 3.
[26] In David Miller's view, there are three conditions that must be satisfied for a subsistence war to be even initially justified. One is 'the responsibility condition', stating that 'liberal and decent peoples are each collectively responsible for the cultural and other features that give rise to inequalities between them'—which, in Miller's view, is rarely satisfied. See David Miller, 'Responsibility and International Inequality in the Law of Peoples', in Rex Martin and David A. Reidy (eds.), *Rawls's Law of Peoples: A Realistic Utopia?* (Oxford: Blackwell, 2006), 191.
[27] The classic articulation of this objection is David Luban's 'Just War and Human Rights'.

recognized by the current *ad bellum* regime, which allows states to wage war in defence of their territorial integrity, yet forbids them from instigating war to prevent systematic oppression of rights in a neighbouring country.[28]

Individualism and Pacifism

If, as *Continuity* requires, wars must satisfy the same conditions for legitimate self-defence that apply to private individuals, and if regular wars of national defence fail in this respect, such wars cannot be morally justified. Proponents should endorse pacifism with respect to what the law regards as 'defensive wars'. Indeed, Norman admits that 'the failure both of utilitarian arguments and of "just war" arguments seems to be pushing us in the direction of a pacifist conclusion'.[29] Rodin concedes likewise: '[i]t might be supposed that the only conclusion we can draw from this result will be a form of pacifism.'[30]

Quite surprisingly, most revisionists do not take this step.[31] They raise creative ideas about how international law and the relevant international institutions should be reformed,[32] but they acknowledge that such reforms might take a long time (if at all accepted).[33] In the meantime, they agree that politicians are allowed to lead their countries into national-defence wars; and that their citizens may enlist, although they will be employed for causes that often do not morally justify war (regular national-defence wars). How could such a position be squared with the individualist view?

Cécile Fabre offers an attempt to defend the accepted *ad bellum* regime and to avoid pacifism. Fabre concedes that not all unjust threats justify lethal measures in defence, but believes that this does not preclude national-defence wars. In so arguing, she distinguishes between a 'narrow just cause' for war and a 'wide just cause'. The former provides its holders with a right to use lethal force against their enemy. The latter provides them with a right to use non-lethal measures only, but

---

[28] See Michael Byers and Simon Chesterman, 'Changing the Rules about Rules? Unilateral Humanitarian Intervention and the Future of International Law', in J. L. Holzgrefe and Robert O. Keohane (eds.), *Humanitarian Intervention: Ethical, Legal and Political Dilemmas* (Cambridge: Cambridge University Press, 2003), 177–203, referred to by Fabre, *Cosmopolitan War*, 132 (who argues that, morally speaking, there might be a right to wage such wars).

[29] Norman, *Ethics, Killing and War*, 207.

[30] Rodin, *War and Self-Defense*, 163.

[31] Fabre, *Cosmopolitan War*, 198, as well as Bradley J. Strawser, 'Walking the Tightrope of Just War', *Analysis* 71/3 (2011), 538–44 and Jeff McMahan, 'Who is Morally Liable to be Killed in War', *Analysis* 71/3 (2011), 544–59.

[32] Jeff McMahan, 'The Morality of War and the Law of War', in David Rodin and Henry Shue (eds.), *Just and Unjust Warriors*. 41–3, Rodin, *War and Self-Defense*, 173–88, Luban, 'Just War and Human Rights', 173–6.

[33] See McMahan, 'The Morality of War and the Ethics of War', 41–3, as well as Rodin, *War and Self-Defense*, 173–88.

if their enemy uses lethal force in response, they acquire a narrow just cause and may use lethal force in defence. Thus:

> Suppose that A's [Aggressor's] forces bloodlessly invade a tiny empty island over which V [Victim] claims sovereignty, but which is located many thousands of miles away from V's mainland. The invasion alone does not provide V with a narrow just cause for sending its combatants to kill combatantsA [combatants of Aggressor]. This is because the violation of this particular right of V's does not in this case undermine the prospects of V's members for a minimally decent life. However, A's act of aggression does provide V with a wide just cause for using non-lethal force against combatantsA. If combatantsA then pose a lethal threat to combatantsV, the latter have a narrow just cause for killing or seriously maiming them.[34]

All the more so if the aggression is directed not against some remote island, but against Victim's mainland. In this case, combatantsV would have a narrow just cause to use lethal force against combatantsA insofar as they 'have very good reasons to believe that A's regime, if unopposed and thus victorious, would commit further violations of their fundamental human rights'.[35] Thus, in Fabre's view, attacked countries always have a broad just cause for war (which translates very quickly into a narrow just cause); and, in most cases, they have a narrow just cause immediately when their enemy crosses the border.

This is an intriguing proposal, but we do not think it works. What combatantsV would be doing in the remote island case is provoking combatantsA to use lethal force so that they, combatantsV, could then fire back—for the sake of ensuring Victim's continued sovereignty over the island. But if mere sovereignty is not valuable enough to justify war, how could such provocation be morally permissible? With so little to gain (morally) by maintaining the island and so much to lose in terms of human life and rights violations, surely proportionality should require retreat rather than provocation and war.

McMahan offers a different response. In his view, although the expected harm to each individual from losing his or her right to determine his or her collective way of life together with other members of the nation is not that substantial, the sum of all these individual harms certainly is; hence, going to war to prevent this move is not disproportionate. A relatively minor harm to, say, ten million citizens, adds up to a serious harm that justifies the use of coordinated lethal force in the form of war—when such use of force amounts to the last resort.[36]

The problem with this response is that it works both ways. Suppose that Country A has only one million citizens and a pretty efficient army. It is attacked

---

[34] Fabre, *Cosmopolitan War*, 70.   [35] Fabre, *Cosmopolitan War*, 70.
[36] Jeff McMahan, 'What Rights May be Defended By Means of War?' in Fabre and Lazar (eds.), *The Morality of Defensive War*, 131–47. See also Frowe, *Defensive Killing*, 139–45.

by Country B, which has ten million citizens and a mediocre army. If Country A goes to war, the number of people who would suffer significant harm (of course, not only on B's side but also on A's side)—many of them truly innocent[37]—is far greater than the number of people who would benefit from their state not having to give up political sovereignty. But the above argument would imply that Country A would have no right to go to war against Country B, which is hard to accept. More generally, as Jeremy Davis explains,[38] that would make the right to wage war dependent on the relative size of the rival parties, giving large states such a right in cases in which it would be denied of small ones.[39] This is absurd. It means that in those cases in which defensive war seems most appropriate—such as when some big country or empire seeks to swallow up some small country—the latter would not have a right to wage war to defend itself.

Other proposals have been made to justify the accepted *ad bellum* regime,[40] but the cases presented above adequately justify scepticism about their prospects. Given the wholesale bloodshed that is typically involved in war, it seems doubtful that individualists, with their uncompromising commitment to individual morality, have the resources to render it justified. They might manage to do so in some exceptional cases, but Article 51 licenses much more: it grants wide scope to the right of states to wage war when their territorial integrity is threatened.

The fact that the individualist logic rules out wars of national defence does not in itself constitute a rebuttal of *Individualism*, until independent arguments are provided to show the implausibility of pacifism. But since many individualists explicitly *reject* pacifism, they owe us an explanation for how this is compatible with their essential premises—an explanation they fail to provide.

With regard to the over-restrictiveness of the current *ad bellum* regime, *Individualism*'s standard response is that 'in principle' it supports justice-promoting wars of the sorts mentioned, but in most cases practical considerations rule them out. We will return to this distinction between 'in principle' and practical considerations later in the book.

Before moving to issues regarding *jus in bello*, let us take stock of what we have done so far. We introduced the individualist view of morality, which assumes the robustness of human rights; their unaffectedness by the social roles that people happen to occupy or by the social/political affiliations that they have. Next, we

---

[37] Here and elsewhere in the book, when we refer to people as 'innocent' we mean innocent *in the relevant sense*; namely, in the sense of having done nothing by virtue of which they might have lost their right to life vis-à-vis a specific attacker. Typically, this means that they are not responsible for the forced choice between their own lives and that of the attacker. In other senses, such people might not be innocent at all; they might be morally responsible for all kinds of crimes and evils.

[38] Jeremy Davis, 'Toward a Non-Reductionist National Defense', Stockholm Centre for the Ethics of War, Graduate Reading Retreat, Bergamo, Italy, 2015.

[39] See also Daniel Statman, 'Supreme Emergencies Revisited', *Ethics* 117 (2006), 61.

[40] See, for instance, Biggar, *In Defence of War* (New York: Oxford University Press, 2013), Chapter 5 and Anne Stilz, 'Territorial Rights and National Defense' in Fabre and Lazar (eds.), *The Morality of Defensive War*, 203–29.

showed the apparent incompatibility of *Individualism* with traditional just war theory and with the UN Charter. We demonstrated that from an individualist point of view, the Charter is at once too permissive and too restrictive. It is too permissive in the blanket permission it grants states to wage war against threats to their sovereignty and to their territorial integrity. And it is too restrictive in its blanket prohibition against wars for the protection of individuals from violations of their fundamental rights; the right to be free from oppression and discrimination, and the right to live a minimally decent life.

Individualists try to avoid both types of implications. They object to pacifism, on the one hand,[41] and shy away from recommending subsistence wars in reality, on the other.[42] We do not presume to have dealt with all the arguments that they present to justify these moves. We hope that what we have said is enough to weaken *Individualism* and indicate the need for an alternative theoretical framework. Before we turn to developing such a framework, we turn to the criticism offered by *Individualism* against the traditional view regarding *jus in bello* and to its own suggestions with regard to the basic distinctions embedded there.

1.5 *Individualism* and the Moral Rules for Conducting Warfare

Individualist Criticism of Traditional *Jus in Bello*

According to the traditional view, combatants can fight justly in an unjust war. As long as they follow the accepted *in bello* rules, they violate no right their victims hold against them—even if the war they are fighting is illegal and immoral. Individualists find this conception very implausible. As David Rodin puts it, 'the just war doctrine is committed to the seemingly paradoxical position that the war taken as a whole is a crime, yet that each of the individual acts which together constitute the aggressive war is entirely lawful'.[43] Paradoxically, by their very aggression, Unjust Combatants acquire the right to kill enemy combatants who—merely by being members of the attacked country—have done nothing by virtue of which they could be said to have lost their right not to be killed.

A possible response would rely on the idea that when two innocent attackers threaten each other, each has a right to kill the other in self-defence (if that is the

---

[41] Fabre, *Cosmopolitan War*, 198, McMahan, 'Who is Morally Liable to be Killed in War', 547, Rodin, *War and Self-Defense*, 197–9.

[42] As Fabre points out, 'in the light of the foregoing considerations, one might think that a subsistence war could never be just—perhaps almost by definition. But that would be too quick. For although such wars might usually succeed only at the cost of a non-negotiable principle of a just conflict, such as the principle of non-combatant immunity, the door remains open, on principle, for deeming just a subsistence war fought in accordance with that principle' (*Cosmopolitan War*, 92).

[43] Rodin, *War and Self-Defense*, 167.

only way for each to save her life).⁴⁴ Add to this the premise that combatants are for the most part innocent with regard to their participation in war (they are too young, brainwashed by their country's propaganda, and under strong pressure from their superiors and peers to comply with the orders they get), and what follows is that combatants of both sides have a right to shoot at each other. But this move is not very convincing. First, most theories of the right to self-defence reject the idea that under these circumstances both sides have a right to attack each other.⁴⁵ Second, combatants often pose no threat at all to their enemy—for instance, because they are assigned to some negligible, administrative task, or because at the time of the attack they are asleep and thus (as this line of reasoning goes) they are not liable to defensive killing—yet the current *in bello* code does legitimize killing them. Third, the assumption that all combatants are equal in terms of their (lack of) responsibility for the war fought by their country, and therefore morally equal in their rights, is overstated.⁴⁶ Many of them bear at least some responsibility for their participation in the war and for the actions they take in its course. Moreover, if combatants were completely innocent with respect to the killing of enemy combatants, it is a bit puzzling that they should be considered *fully* responsible when it comes to the (intentional) killing of *non*combatants, though the pressures on them to obey their superiors and behave like their peers are the same in both cases. In other words, if combatants have the epistemic and motivational resources to resist orders to commit war crimes, they cannot be so ignorant and weak when it comes to orders to participate in unjust wars. Thus, the traditional moral equality of combatants cannot be based on their equal innocence with respect to the threat that they pose to the other side.

Revisionists infer from the last point that Unjust Combatants are at least minimally culpable for the immoral killing that they bring about. They are like the criminal in the following example⁴⁷:

ROBBERY: During a robbery, the bank guard tries to reach for his gun. Realizing that this was the guard's intention, the thief shoots the guard and kills him. The thief was coerced to go to the robbery by the head of the gang, who had threatened him with severe bodily harm. Initially, the thief had no murderous

---

⁴⁴ For the free competition intuition with respect to cases in which individuals blamelessly attack each other, see Yitzhak Benbaji, 'Culpable Bystanders, Innocent Threats and the Ethics of Self-Defense', *Canadian Journal of Philosophy* 35 (2005), 585–622; Yitzhak Benbaji, 'A Defense of the Traditional War-Convention', *Ethics* 118 (2008, 593–618); Nancy Davis, 'Abortion and Self-Defense', *Philosophy and Public Affairs* 13 (1984), 175–207; and Jonathan Quong, 'Killing in Self-Defense', *Ethics* 119 (2009), 507–37.

⁴⁵ Jeff McMahan, 'Self-Defense and the Problem of the Innocent Attacker', *Ethics* 104 (1994), 252–90; Michael Otsuka, 'Killing the Innocent in Self-Defense', *Philosophy and Public Affairs* 23 (1994), 74–94.

⁴⁶ For a detailed normative analysis of this fact, see Seth Lazar, 'The Responsibility Dilemma for Killing in War: a Review Essay', *Philosophy and Public Affairs* 38/2 (2010), 180–213.

⁴⁷ A modification of an example offered by McMahan, 'The Ethics of Killing in War', 696.

intentions. His killing was purely defensive. Had he not killed the guard, the guard would have killed him.

Needless to say, by trying to stop the robbery, the guard does not lose his right to life; and, in particular, he retains his right not to be attacked by the thief. Hence, killing the guard counts as murder.

Thus, just as the guard and the thief do not have an equal right to shoot at each other in self-defence, Just and Unjust Combatants also lack such an equal right. The traditional view permits the mass killing of innocent human beings—combatants of the just side. This is something that *Individualism* cannot accept.

The second assumption that *Individualism* challenges concerns the categorical difference between combatants and noncombatants. At least in regular wars— namely, wars between *armies* fighting for *states*—only combatants have a right to use lethal force, and they are allowed to do so only against combatants of the other side. War is a game in which only combatants are supposed to take part. But this view is hard to reconcile with *Individualism*.

Let us start with the claim that only combatants have a right to kill in war.[48] Suppose that an aggressor poses an unjust threat to a group of people, only killing him can neutralize the threat, and such killing is proportionate to the threat. Under those conditions, it is hard to see why only some members of the group would have a right to kill him and not others. To be sure, it is not often the case that noncombatants have the opportunity to kill enemy combatants—the war is usually far from home, the noncombatants are not armed, and so on; but when they do, why should such acts be ruled out? If the enemy combatant is liable to be killed because of the unjust threat he poses, what moral difference does it make if he is killed by John, a combatant, or by John's younger brother, Matt, who is a civilian?

Moreover, if a person has a right to self-defence against some aggressor, then third parties usually have a right to intervene on her behalf and protect her from the unjust threat posed against her. But if third parties have such a right, even when none of their own interests are threatened, all the more so with people like Matt who are threatened by the unjust attack like other members of the just country. According to the traditional view, with regard to wars between regular armies, there is a sharp moral distinction between these killings.[49] This is

---

[48] See Haque's discussion in, *Law and Morality at War*, 25. While Haque insists that the law does not prohibit the participation of civilians in war, he concedes that the law treats the killings committed by civilians and those committed by soldiers differently. According to Haque, only soldiers are immune from legal prosecution by a foreign state.

[49] See Yoram Dinstein, *War, Aggression and Self-Defence* (Cambridge, MA: Cambridge University Press, 2001), 27–9. See also George P. Fletcher, *Romantics at War: Glory and Guilt in the Age of Terrorism* (Princeton, NJ: Princeton University Press, 2002), who concludes that the correct result under international law—'as hard as it might be to swallow'—is that somebody like Matt 'is guilty of murder under domestic law' (59).

inconsistent with *Individualism*'s insistence that social and organizational affiliation makes no difference when fundamental rights are at stake.

Turning now to the assumed immunity of noncombatants from deliberate attack, this notion seems incompatible with the paradigm of individual self-defence. Most individualists assume that the basis for liability for defensive killing is responsibility for an unjust threat, rather than merely the very posing of such a threat. But this implies that the more one is morally responsible for an unjust attack, the more one is liable to defensive attack. Since many noncombatants are more morally responsible than 18-year-old combatants for the threat posed by their country—think of politicians, civil servants, public intellectuals, or scientists—they should count as *more* liable to defensive killing than the latter, not less.[50] How could it be, ask individualists, that the former are categorically protected, both morally and legally, from attack, while the latter are free game? If a person is morally responsible for an unjust attack, then she is a legitimate target for defensive action regardless of her social role or political affiliation. If she is not, then she is not such a legitimate target—again, regardless of social role or political affiliation. Such logic leads to a rejection of the two most basic premises of traditional *jus in bello*—that no civilians are legitimate targets of attack and that all combatants are.

Finally, *Individualism* faces difficulty in explaining the permission to harm civilians (if the harm is not excessive) as a side effect of attacks on military targets. This permission seems incompatible with proper respect for those civilians, many of whom are clearly innocent. This difficulty increases when one compares the permission to collaterally harm civilians with the prohibition against intentionally attacking them. While intentionally killing innocent people is, in some views, worse than foreseeably killing them (other things being equal), this does not yield the conclusion that foreseeably (yet unintentionally) killing civilians is *permissible*, even if the killing would standardly be considered as a lesser evil.

The usual response to this difficulty relies on a widespread intuition regarding a well-known trolley case. Consider:

---

[50] Note Mavrodes's comment on this point: 'Now, we should notice carefully that a person may be an enthusiastic supporter of the unjust war and its unjust aims, he may give to it his voice and his vote, he may have done everything in his power to procure it when it was yet but a prospect, now that it is in progress he may contribute to it both his savings and the work which he knows best how to do, and he may avidly hope to share in the unjust gains which will follow if the war is successful. But such a person may clearly be a noncombatant, and (in the sense of the immunity theorists) unquestionably "innocent" of the war. On the other hand, a young man of limited mental ability and almost no education may be drafted, put into uniform, trained for a few weeks, and sent to the front as a replacement in a low-grade unit. He may have no understanding of what the war is about, and no heart for it. He might want nothing more than to go back to his town and the life he led before. But he is "engaged," carrying ammunition, perhaps, or stringing telephone wire or even banging away ineffectually with his rifle. He is without doubt a combatant, and "guilty," a fit subject for intentional slaughter. Is it not clear that "innocence," as used here, leaves out entirely all of the relevant moral considerations – that it has no moral content at all?', George I. Mavrodes, 'Conventions and the Morality of War', *Philosophy and Public Affairs* 4/2 (1975), 122–3.

DIVERSION: You are the driver of a trolley. The trolley rounds a bend, and five track workmen, who have been repairing the lines come into view. You step on the brakes, but, alas, they don't work. Now you suddenly see a spur of track leading off to the right. You can turn the trolley onto it, and thus save the five men on the straight track ahead. Unfortunately, there is one workman on that spur of track. He can no more get off the track in time than the five can, so you will kill him if you turn the trolley onto him.[51]

In this case, the trolley driver knowingly selects a course of action that is predicted to lead to the death of an innocent person; nonetheless, he is permitted (maybe even obliged) to do so. If such killing is permitted in DIVERSION, it can be permitted in war too—namely, in the foreseeable (yet unintentional) killing of civilians.

As for the difference between the latter permission and the outright prohibition against targeting civilians intentionally, that seems to follow the widely accepted difference between DIVERSION and

FAT MAN: A trolley is hurtling down a track towards five people. You are on a bridge under which it will pass, and you can stop it by putting something very heavy in front of it. As it happens, there is a very fat man next to you—your only way to stop the trolley is to push him over the bridge and onto the track, killing him to save five.[52]

Almost everybody agrees that while in DIVERSION it is permissible to save the five by causing the death of the one, in FAT MAN, this is out of the question. The common explanation for this intuition is that by throwing the fat man on the track you would be using him as a means to save the five. In somewhat technical terms, such killing would be 'opportunistic', 'manipulative', or 'exploitative'. This description would not hold true for DIVERSION, in which the victim cannot be said to 'be cast in some role that serves the agent's goal'[53]; hence, standard lesser-evil considerations may guide the driver in his decision.

This explanation appeals to the Doctrine of Double Effect (DDE). The DDE states that 'an act can be impermissible if done with a wrongful intention, even if the same act—or at least an act involving the same physical movements and having the same consequences—would be permissible if done with an acceptable intention'.[54] Thus, the DDE appeals to the intended/foreseen distinction to

---

[51] Judith J. Thomson, 'The Trolley Problem', *Yale Law Journal* 94/6 (1985), 1395–415.
[52] Ibid., 1409.
[53] Warren S. Quinn, 'Actions, Intentions, and Consequences: The Doctrine of Double Effect', *Philosophy and Public Affairs* 18/4 (1989), 349.
[54] Jeff McMahan, 'Intention, Permissibility, Terrorism, and War', *Philosophical Perspectives* 23 (2009), 345–6.

differentiate between FAT MAN and DIVERSION. In the former case, the killing of the victim is intended, while in the latter cases the death is foreseen but unintended. The DDE states, therefore, that the killing in FAT MAN is harder to justify than the killing in DIVERSION.

Applying these distinctions to war, individualists might argue that usually, intentionally killing civilians is killing them manipulatively, like in FAT MAN, hence forbidden; whereas killing them unintentionally, or collaterally, involves no manipulation and is (hence) permissible.

However, prominent philosophers, most notably Thomson, find the DIVERSION/FAT MAN distinction morally insignificant. They argue that the incidental killing in the former case is impermissible as well.[55] Famously, Thomson also denies the relevance of intentions, and more generally mental states, to permissibility.[56]

Furthermore, even if intentions are relevant to permissibility, it is unclear why the mental state of the driver in DIVERSION is (metaphysically and normatively) so different from the mental state of the agent in FAT MAN. The agent in FAT MAN intends the death of the fat man—although she regrets it, tries to avoid it, and generally does not want it. The death of the FAT MAN is intended by virtue of the agent's preference; she prefers to kill him as a means to save the five, rather than to let the trolley kill them. The problem is that the agent in DIVERSION has a similar preference; she also prefers to cause the death of one, rather than to let the trolley kill five. Why, then, is the death of the sidetrack man considered unintended but that of the fat man intended? And even if ordinary language does draw a distinction between these agents, why is it morally significant?

Finally, even if the killing in FAT MAN is impermissible (on grounds of being manipulative), this provides only a partial explanation for the prohibition against intentionally killing innocents, since intentional killing can be non-manipulative. Consider:

ALCOVE: Ralph is being chased by a pursuer who wants to kill him. To save his life, Ralph must hide in some alcove. Unfortunately, the alcove is occupied by some homeless woman who refuses to move. The only way Ralph can get her out is by killing her.

Although Ralph would not be using the woman as a means for his survival, killing her seems intentional. It seems impermissible, as well. Cases like ALCOVE are not that rare in war. Here is an example:

---

[55] Judith J. Thomson, 'Turning the Trolley', *Philosophy and Public Affairs* 36 (2008), 359–74.
[56] Judith J, Thomson, 'Self Defense', *Philosophy and Public Affairs* 20 (1991), 283–310. See also David Rodin, 'Terrorism without Intention', *Ethics* 114 (2004), 752–71.

OBSTACLE1: A team of soldiers is trying to reach an important enemy facility located in enemy territory. Two children discover their whereabouts. If these children are not killed, they will innocently report the team, thereby jeopardizing the mission (and the lives of the soldiers).

Killing the children is intuitively wrong and clearly prohibited by *Civilian Immunity*. This intuitive verdict will not change even if the killing is unintended, as in

OBSTACLE2: A rescuer can reach a switch with which she can stop a moving trolley that would otherwise kill five innocent people. However, in order to reach the switch, the rescuer must run over and kill two children who are in her way.

Again, why are these cases of intentional and unintentional killings so different from the incidental killing cases, to which the *Immunity* prohibition does not apply?

Relying on DIVERSION to rebut the intuitive objection against the permission to foreseeably kill the innocent is problematic for another reason. Let us grant that in some cases, the permission to kill is determined by standard lesser-evil calculations. You may prefer to rescue five people, even if this involves killing another. However, in the context of war, there is no basis for assuming that standard lesser-evil considerations can justify attacks that involve the collateral killing of enemy civilians. It is simply not the case that the overall number of innocent lives saved by such attacks is always larger than the number of innocent lives lost; definitely not *significantly* larger (a 5:1 ratio), like in DIVERSION. Thus, the permission in DIVERSION is not adequate to dispel the sense of moral bewilderment generated by the broad right to foreseeably (yet unintentionally) bring about an innocent death in the course of war.

In sum, *Individualism* is at once more permissive and more restrictive than the accepted *in bello* code (just as it is with regard to the accepted *ad bellum* code). It is more permissive because it is forced to include culpable civilians (civilians who are responsible for an unjust threat) within the category of legitimate targets. It is more restrictive because it denies the right of Unjust Combatants to kill Just Combatants, rejects the presumption that all combatants are legitimate targets, and has difficulties in granting permission to foreseeably though unintentionally kill civilians.

These implications of *Individualism* are quite alarming. The permissive aspect is alarming as it opens the door to total war—war that blurs the boundaries between combatants and noncombatants, between battlefields and residential areas. The restrictive aspects are alarming because they undermine the practical possibility of waging a just war, thus paving the way to pacifism. If many combatants are illegitimate targets, it is hard to see how wars could be fought in a way that is both effective and morally justified.

These implications are alarming even for individualists, who try either to evade them or at least to soften them in a way that would render compliance with the traditional *in bello* code morally acceptable. Ultimately, after this rigid criticism, individualists seem to resubscribe to the basic tenets of the *in bello* code—at least until some dramatic progress is made in the international regime.[57] In Lazar's estimation, McMahan's later work 'is distinctive less for its opposition to Walzer, than for its compromises with its guiding intuitions'.[58] More recently, James Pattison argues that:

> [o]nce we look at the remarks of revisionists when they *do* offer non-ideal, applied accounts, we can see that the leading traditionalists and revisionists are in fact much closer in their substantive accounts. Most notably, despite his challenges to the principles of *Civilian Immunity* and the equality of combatants, McMahan defends maintaining current international humanitarian law, at least in the short term... Thus, Walzer notes that, in practice, McMahan more or less adopts the same view as he does. He thus claims that "[o]ur disagreement, at the end, may be only terminological."[59]

Can Individualism Resubscribe to the Accepted In Bello Code?

(I.) *Individualism* tries to accommodate at least three aspects of traditional just war theory without betraying its fundamental principles. The first is the immunity from legal persecution conferred on Unjust Combatants. Unjust Combatants have no moral right to kill Just Combatants. At face value, this assumption should have straightforward legal implications; namely, a moral duty to bring Unjust Combatants to trial and make them pay for their assumed criminal activity. However, *Individualism* denies that legal systems face such a duty.

One reason might be practical: bringing all Unjust Combatants to court would be practically impossible. But this feasibility-based reasoning does not support going to the other extreme and accepting that *all* combatants should be immune from legal measures against them (unless they commit war crimes). If a large group conspires to commit a mass murder of innocents—which is precisely the case with unjust wars, from the individualist perspective—then even if it is impractical to punish all group members, surely many of them can and should be punished; in particular the leaders and those whose hands are especially tainted with innocent blood.

---

[57] See in particular McMahan, 'The Ethics of Killing in War', 729–33.
[58] See Lazar, 'The Responsibility Dilemma for Killing in War: a Review Essay', 181.
[59] James Pattison, 'The Case for the Nonideal Morality of War: Beyond Revisionism versus Traditionalism in Just War Theory', *Political Theory* 46/2 (2018), 256–7, referring to Michael Walzer, 'Response to McMahan's Paper', *Philosophia* 34 (2006), 43–5.

Individualists might explain this legal immunity by appealing to the limited access of soldiers to the relevant knowledge, to their young age, and to the strong pressure from their officers and peers—all of which make Unjust Combatants fully excused for their participation in the unjust wars initiated by their countries. Alas, such a blanket exemption from responsibility is arbitrary and would be unacceptable in any domestic self-defence context. If a bunch of criminals conspires to carry out the mass murder of innocent people, then although some of them might turn out to be fully excused for their participation in this criminal plot, no legal system would assume a priori that *all* of them are.

McMahan suggests a different explanation. If combatants knew that they might face trial when the war was over, they would have weaker motivation to end the war than if they knew that as long as they fought in accordance with the *in bello* requirements they would not be subject to any legal measures. Hence, he argues, for the sake of shortening wars, it is better not to criminalize wrongful killing carried out by Unjust Combatants.[60]

This line of argument fails once we look at its domestic analogy; if murderers knew that they faced life imprisonment (or worse, capital punishment) for the crimes that they had already committed, they would have no incentive to stop their criminal activity; hence, their crimes ought not to be punished. It seems incomprehensible that if this were true, murderers would be let off the hook and face no charges for their crimes.

(II.) A second element in the traditional *in bello* code that individualists tend to adopt is the blanket permission to kill Unjust Combatants. How could individualists accept this permission, given that many enemy combatants are not even minimally responsible for the unjust war in which they participate? A standard individualist response is to lower the bar for the level of responsibility necessary to make one liable to defensive killing. Notwithstanding the limited access of combatants to the relevant knowledge and the social pressures put on them to comply, they are responsible for the threat they pose. Exempting them from all responsibility (as if they were psychopaths or people operating under the influence of drugs) will not work. Soldiers know what they are doing and can be said to be the *agents* of their actions. Assumedly, this agency is sufficient to make them liable to defensive killing. Thus, since by definition, Just Combatants are completely innocent (in the relevant sense), and since virtually all Unjust Combatants are at least minimally responsible for their participation in the unjust war waged by their country, it is the latter who should bear the cost, not the former.

This reasoning comprises the basis of an argument for asymmetrical permission to kill combatants, which Seth Lazar puts as follows:

---

[60] See Jeff McMahan, 'Killing in War: A Reply to Walzer', *Philosophia* 34 (2006), 50.

1. Barring a lesser-evil justification, in war Just Combatants may permissibly intentionally kill only people who are liable to be intentionally killed.
2. An individual is liable to be intentionally killed if and only if he is
   a. sufficiently responsible for an unjust threat
   b. that is sufficiently serious to make killing him proportionate
   c. and killing him is necessary to avert that threat.
3. Lesser-evil justifications for intentional killing in war are rare enough to be morally insignificant exceptions.
4. Besides morally insignificant exceptions, in war all Unjust Combatants satisfy the criteria in premise 2, and all Unjust Noncombatants do not.
5. Therefore, besides morally insignificant exceptions, Just Combatants may permissibly intentionally kill only Unjust Combatants in war.[61]

In particular, McMahan insists that the sort of minimal responsibility that is required to make a person liable to defensive killing is not culpability:

> If a morally responsible agent—that is, an agent with the capacity for autonomous deliberation and action—creates an unjust threat through voluntary action that is wrongful but fully excused, she is to some extent responsible for that threat even though she is not blamable.[62]

Let's call this view

Agent Responsibility: The responsibility required for a person to be liable to defensive killing for posing a threat to Victim is defined in terms of his/her having the capacity for autonomous deliberation and action.

Applying this principle to war demonstrates how, despite the fact that many combatants are fully excused for their participation in an unjust war, they can be morally responsible for such participation and hence liable to defensive killing.

This switch to *Agent Responsibility* fails.[63] Even if *Agent Responsibility* is accepted, it falls short of substantiating the right to indiscriminately attack enemy combatants. Consider the following example:

SLEEPING COMBATANTS: An unjust war is initiated with an attack carried out by an elite unit of Unjust Combatants. The preparations for the military

---

[61] See Seth Lazar, 'Liability and the Ethics of War: A Response to Strawser and McMahan', in Christian Coons and Michael Weber (eds.), *The Ethics of Self Defense* (Oxford: Oxford University Press, 2011), 292–304.

[62] McMahan, 'The Ethics of Killing in War', 723.

[63] In substantiating this argument, we rely on Yitzhak Benbaji, 'The Responsibility of Soldiers and the Ethics of Killing in War', *The Philosophical Quarterly* 57 (2007), 558–73.

campaign were highly confidential. Very few Unjust Combatants are aware of them. Immediately after the surprise attack is launched, Just Combatants respond by attacking sleeping enemy combatants. These combatants of the unjust side contributed nothing to the unjust attack, knew nothing about it in advance, and are not responsible for the aggression initiated by their country in any other way.

As the sleeping combatants did nothing to make themselves liable to the killing, Just Combatants ought to avoid targeting them. This result does not change even if *Agent Responsibility* is assumed, because sleeping combatants are not agents of any action, positive or negative. What follows is that not all Unjust Combatants are legitimate targets.

McMahan argues that the liability of the sleeping soldiers is grounded in their joining the army, even if at the time they joined, joining the army was permissible. But this is hard to accept. To advance such an argument, McMahan proposes the following example:

CONSCIENTIOUS DRIVER: A person keeps his car well maintained and always drives cautiously and alertly. On one occasion, however, freak circumstances cause the car to go out of control. It veers in the direction of a pedestrian whom it will kill unless she blows it up by using one of the explosive devices with which pedestrians in philosophical examples are typically equipped.[64]

May the pedestrian blow up the driver in order to save her life? McMahan believes that she may and assumes that most people would agree with this position. How could this judgment be justified? Here is his answer:

> What makes him liable is that, as a morally responsible agent, he voluntarily chose to set a couple of tons of steel rolling as a means of pursuing his ends, knowing that this would involve a tiny risk that he would lose control of this dangerous object that he had set in motion, thereby imperiling the lives of the innocent... [He is] liable because he voluntarily engaged in a risk-imposing activity and is responsible for the consequences when the risks he imposed eventuate in harms.[65]

The moral intuition that underlies this judgment is the same as that which underlies the acceptance of option luck in luck egalitarianism. According to luck egalitarianism, the distribution of goods should not be influenced by brute luck; i.e., by factors over which we have no control. But once the initial distribution is just, there is nothing unfair in gaps created by the way the players choose to use

---

[64] Jeff McMahan, 'The Basis of Moral Liability to Defensive Killing', *Philosophical Issues* 15 (2005), 393.
[65] Ibid., 394.

their resources, by the way they 'play their cards'. Similarly in the present context, if a person voluntarily chooses to engage in an activity that imposes risks on others, she thereby undertakes the risk that she may have to bear the cost of doing so. Since, at t1, the agent voluntarily chose to engage in a risk-imposing activity, she is liable to defensive measures for a threat she imposes at t2, even though at t2 she is not even agent-responsible for this threat.

Returning to war. One might assume, on the basis of this argument, that all Unjust Combatants are legitimate targets—including those in SLEEPING COMBATANTS. By joining the military, combatants undertake the risk that they will be deployed for all kinds of missions, including unjust ones. Therefore, they make legitimate targets for defensive killing by the innocent people that they endanger (i.e., combatants of the just side), exactly as the driver is a legitimate target for such killing by the pedestrian. By contrast, no such undertaking can be ascribed to civilians; hence the categorical difference between civilians and combatants.

We believe, however, that this argument fails. Consider the following:

CAREFUL PHYSICIAN: A careful physician caused an unbearably painful kidney ailment by giving her patient some drug. Had the physician known the consequences in advance, she would have avoided the treatment, for the life of the patient is now not worth living. Still, the physician's decision to prescribe the drug was careful and well informed. The risk in consuming the drug was infinitesimal and hence worth taking, especially as the drug was the only chance the patient had to save his life. It so happens that by using the physician's body in some very harmful way the patient can be cured.

The option luck logic should lead one to conclude that the physician's body may be used this way. When choosing to work as a doctor, the physician knew that she might be imposing risks on patients by offering them treatment that would make their situation worse. In the US, a little over 900,000 doctors[66] cause about 100,000 deaths every year as a result of medical errors,[67] which means—again very roughly—that on average each of them kills a patient every nine years. Fewer are killed (on average) every year in car accidents.[68] Therefore, if the driver in CONSCIENTIOUS DRIVER is liable to defensive killing because he assumedly undertook the risk of becoming liable to defensive measures as a result of posing a threat to the lives of others, all the more so with the physician in CAREFUL

---

[66] See Aaron Young, Humayun J. Chaudhry, Xiaomei Pei, Katie Halbesleben, Donald H. Polk, and Michael Dugan, 'A Census of Actively Licensed Physicians in the United States', *Journal of Medical Regulation* 101/2 (2015), 8–23.

[67] For different estimates, see Ronen Avraham, 'Private Regulation', *Harvard Journal of Law and Public Policy* 543 (2011), 549, fn. 13.

[68] For statistics, see United States Department of Transportation National Highway Traffic Safety Administration, Fatality Analysis Reporting System (FARC) Encyclopedia, Accessed 27/02/2017, available at https://www-fars.nhtsa.dot.gov/Main/index.aspx.

PHYSICIAN. But that is clearly absurd. The patient has no right to get rid of the pains caused by the kidney ailment by coercively using the physician's body.

Proponents of the responsibility account might argue that the physician's immunity to defensive attack is an exception based on two considerations. First, physicians intend to act in the interest of the patient; and second, patients consent or would have consented to be put at risk by their physicians. Yet this response would imply that although the risk-imposing driver is liable to defensive killing, an ambulance driver who intends to save people, but actually poses a threat to them, is not—a result that is hard to accept.

The dilemma faced by the responsibility account is simple. Prospectively, risk-imposing activity is either justified or not. Risks might or might not be worth taking. We are strongly inclined to think that the action of the careful physician is morally justified, despite its foreseeable bad consequences. If so, we submit, the driver's risk-imposing activity is justified as well. There is no reason to distinguish between the careful physician and the risk-imposing driver just because the latter acts out of self-interest.

The second reason for which CAREFUL PHYSICIAN might be thought to be an exception, the putative consent of the patient, is even stranger than the first. It implies that had the patient not consented (if he was in a coma, say), considerations of justice would allow some third party to use the physician's body coercively and harmfully to cure the patient, on the ground that the physician had harmed his patient. But that is absurd. Patient consent is of the utmost importance; by insisting on it, policymakers urge patients to make sure that the risk imposed on them is worth taking. But the well-informed patient would acknowledge that if something goes wrong, she has no justice-based reason for shifting a foreseeable harm to the careful physician, despite the fact that the physician has caused it.

To conclude this point, the mere fact that one voluntarily engages in a dangerous activity like driving or working as physician is insufficient to make one liable to defensive killing in case such killing is necessary to prevent the danger from materializing. The same—finally—applies to combatants. Being a combatant carries the risk of causing unjust death; but that status alone is insufficient to grant her enemy a blanket right to kill her even when she poses no threat—for example, when she is asleep.[69]

That is not to deny that undertaking a social role makes a moral difference; it actually makes a *huge* difference. However, as we show in the next chapter, the

---

[69] For an attempt to justify the permission to kill combatants on the basis of a collectivist understanding of war, see Zohar, 'Collective War and Individualistic Ethics' and 'Should the Naked Soldier be Spared? A Review Essay of Larry May, *War Crimes and Just War*', *Social Theory and Practice* 34/4 (2008), 623–34. See especially note 21 on p. 633 with regard to the identity of the relevant collective ('Initially I tended to identify the collective entity as "the nation"; I am now inclined instead to focus on the army as a collective agent').

meaning of this undertaking and its relation to basic human rights are different from those assumed by the revisionist argument that McMahan advances. Properly understood, they lead to a very different moral picture than that supposed by *Individualism*.

(III.) The third aspect of the traditional war convention that individualists by and large accept is *Civilian Immunity*. However, the move to *Agent Responsibility* (further) undermines the fundamental distinction between combatants and noncombatants. Seth Lazar has developed this point well. It follows from the individualist argument offered above that a soldier is liable to intentional killing if killing her comprises a necessary and proportionate means to avert an unjustified threat, T, and if the soldier is sufficiently responsible for T. The *Agent Responsibility* move implies that if necessary to avert a wrongful threat, T, intentionally killing a soldier is permissible even if the degree to which the soldier is responsible for T is low.[70] However, if the responsibility bar is lowered to include all combatants, then many noncombatants will be liable too: they contribute to the war by 'paying taxes that fund the war, supplying military necessities, voting, supporting the war, giving it legitimacy ... bringing up and motivating the sons and daughters who do the fighting'.[71] On the other hand, if to become liable to defensive killing, aggressors must be culpable for the wrongful threat, then many combatants would be spared on grounds of being too young, brainwashed, and under pressure by their commanders and peers to comply. However you play it, the principles of individual morality enshrined in *Individualism* are incompatible with the distinction between combatants and noncombatants, which is the most important principle of the accepted *in bello* regime.

Lazar addresses another important necessity-based argument for immunity: the actual contribution of most noncombatants to the threat posed by their countries is very low; hence, killing them usually makes no contribution to the elimination of the (perceived) unjust threat. But this argument provides only partial relief. First, even regarding combatants, the actual contribution of many of them to the military effort is pretty low; therefore, the military benefit ensuing from killing them is at best marginal. Second, occasionally, attacking noncombatants is more effective in advancing victory than the argument assumes. An extreme example would be the attacks on Hiroshima and Nagasaki. There are probably other cases in which deliberate attacks on civilians helped the attackers to achieve their goals (typically some form of national liberation).[72]

Third, even if directly killing civilians is usually ineffective and therefore ruled out, this is not the case with killing them as a side effect of targeting legitimate military targets. We have assumed so far that if such targeting brings about the

---

[70] See Seth Lazar, *Sparing Civilians* (Oxford: Oxford University Press, 2015), 9.
[71] Lazar, 'The Responsibility Dilemma for Killing in War: A Review Essay', 192.
[72] Lazar, 'The Responsibility Dilemma', 209–10.

death of civilians, it requires special justification, because the latter are innocent people who have done nothing by virtue of which they might have lost their right to life; hence the requirement that such collateral killing of civilians be proportionate (to the military target). But if the responsibility bar is lower in the way explained above, the innocence (again, in the relevant sense) of such civilian casualties can no longer be assumed. Many of them are definitely more responsible for the unjust war conducted by their country than civilians (and combatants) of the just side. In the forced choice between their own lives and the lives of people on the just side, the lives of the latter should take precedence. This seems to imply that they cannot complain when killed as a side effect of an attack on a legitimate military target, even when such killing is clearly disproportionate or 'excessive' (to use the Geneva Convention term[73]).

1.6 Conclusion

The individualist critique has mercilessly exposed the naiveté of the view that the principles governing the morality of war are just an extension of the right to individual self-defence. This exposure has undermined virtually all the principles underlying both the *ad bellum* and the *in bello* codes: the right to wage war in defence of political borders; the prohibition against targeting civilians; the right to kill civilians as a side effect; the almost unlimited right to kill enemy soldiers regardless of the justness of their cause; and the right to kill combatants regardless of their material contribution to or moral responsibility for the threat posed by their country.

What follows is that according to *Individualism*, much (maybe most) of the killing that takes place in war is morally unjustified. If one turns out to be fighting for an unjust cause—and there is a non-negligible chance that one is—then *all* the killing one brings about is unjustified, of Just Combatants and Just Noncombatants (whether targeted directly or as a side effect) alike.[74] But even if one turns out to be fighting for a just cause, one would be killing many people who are not legitimate targets for defensive killing; namely, many Unjust Combatants, as well as many Unjust Civilians who are killed collaterally. Given the powerful presumption against killing the innocent, contingent pacifism seems the only

---

[73] See Article 51(5)(b) of the 1977 Additional Protocol I, which prohibits attacks that 'may be expected to cause incidental loss of civilian life, injury to civilians, damage to civilian objects, or a combination thereof, which would be excessive in relation to the concrete and direct military advantage anticipated'.

[74] This is a bit of an overstatement, because revisionists concede that there may be circumstances under which even Unjust Combatants would have a right to kill their adversaries; for instance, if the latter commit war crimes, see Jeff McMahan, *Killing in War* (Oxford: Oxford University Press, 2009), 16–17. See also Victor Tadros, 'Unjust Wars Worth Fighting For', *Journal of Practical Ethics* 4/1 (2016), 52–78.

respectable response. What individualists should have said is: 'Sorry folks, this mass killing and maiming called war is simply incompatible with proper respect for humanity; hence, you must refrain from it.'

But most individualists refuse to take this route. They propose various reforms in the laws of war and in the international institutions that regulate them; but until these reforms are implemented (if they ever are), they accept the moral authority of the current arrangements. 'It is dangerous', says McMahan, 'to tamper with rules that already command a high degree of allegiance. The stakes are too high to allow for much experimentation with alternatives.'[75] This response might have worked if the stakes were not as high as they are. The problem is that if the individualist arguments are sound, the war convention permits a violation of rights so massive that nobody with a serious commitment to human rights could accept it.

Individualism cannot ground its contingent acceptance of the current war convention (i.e., until it is reformed) on consequentialist, lesser-evil considerations because such logic would betray its commitment to human rights. If *Individualism* is followed to its logical end, it is unable to offer justification for most of the rules that govern the ethics of war, on both the *ad bellum* and the *in bello* levels. The naiveté of traditional just war theory alongside the failure of *Individualism* to offer a viable substitute to it emphasize the need for an alternative. This is the challenge that we try to face in the rest of the book.

---

[75] McMahan, 'The Ethics of Killing in War', 731.

2
Foundations of a Non-Individualist Morality

2.1 Introduction

In the previous chapter, we saw how the individualist approach to the ethics of war is informed by a general view on the nature of morality—a view that emphasizes the stability of basic rights. *Individualism* states that the distribution of these rights is unaffected by the social role or affiliation of the relevant parties. Individualists who take rights seriously must acknowledge this robustness. *Individualism* must shape the moral domain in accordance with the deep moral relations that bind all people as bearers of natural rights, not in accordance with morally superficial relations defined by contingent social circumstances.

The individualist call for taking rights seriously, especially in war, is well taken. However, we posit that social roles are not at all superficial; rather, they play a crucial role in determining moral relations with others. The purpose of this chapter is to develop this idea and to elaborate on its theoretical commitments and underpinnings. This will serve as the theoretical background to the contractarian account on the ethics of war that we offer in the following chapters. Note, though, that in this chapter we do not offer a full defence of the theoretical approach we present. We sketch initial arguments here, and in later chapters we demonstrate the explanatory force of this approach in substantiating the morality of the traditional war convention.

We start, in Section 2.2, by arguing for the 'moral effectiveness' of some social rules. We offer a series of examples that supports the idea that the existence of social rules affects the distribution of moral rights and duties among members of society. In Section 2.3 we explicate which conditions are met by morally effective rules. In Section 2.4 we focus on a particular function of social rules; namely, on how they constitute social *roles*. If social rules that institute roles are morally effective, so we argue, they entail a large-scale distribution of rights and duties among role holders and others. In Sections 2.5–2.7 we elaborate on a further condition for the moral effectiveness of social arrangements in light of an important objection. We conclude by elucidating the differences between the version of contractarianism that we sketch in this chapter and rule-consequentialism.

2.2 Social Rules and Morality

The basic idea we put forward in what follows is this: in many circumstances, social rules determine fundamental moral rights and duties and thereby shape—or reshape—the moral landscape. As Hart's practice theory of rules states, the social rules to which a group is subject are constituted by 'both patterns of conduct regularly followed by most members of the group and a distinctive normative attitude to such patterns of conduct'. Hart called this attitude to social rules 'acceptance'. It 'consists in the standing disposition of individuals to take such patterns of conduct both as guides to their own future conduct and as standards of criticism which may legitimate demands and various forms of pressure for conformity'. In other words, individuals subject themselves to rules by taking the 'internal point of view' with respect to them; they treat them as 'content independent' reasons to act. From the external point of view, an observer would realize that a participant in such practice 'accepts the rules as guides to conduct and as standards of criticism'.[1] Social rules underlie a 'social practice'. The social practice that social rules underlie consists of 'the fact that the members of a given group generally behave in a certain way, have certain expectations and intentions, and accept certain principles as norms'.[2]

Therefore, as an alternative to the robustness of rights entailed by *Individualism*, we propose that under given conditions, a person's acceptance of social rules can be seen as his/her granting consent to be governed by them, hence as a waiver of rights. This leads to:

Social Distribution: Under specified conditions, social rules partly determine the distribution of moral rights and duties.[3]

The understanding that social rules determine a distribution of rights and duties is, of course, trivial within a Hobbesian view of morality. According to Hobbes, outside organized ('civic') society, people have no moral duties towards their fellow humans and may do whatever is needed to advance their own interests. But this view is widely dismissed in moral philosophy. Therefore, our starting point is Lockean. We assume the existence of 'natural' rights and duties; namely, rights and duties whose validity does not depend on any kind of social agreement.

---

[1] H. A. L. Hart, *The Concept of Law* (Oxford: Oxford University Press, 1994), 255–6.
[2] Thomas Scanlon, *What We Owe to Each Other* (Cambridge, MA: Harvard University Press, 1998), 295.
[3] For the concept of social rules, see Hart's discussion in *The Concept of Law*, 55–7, and Chapter 5. For a principled argument for the moral relevance of social norms, see Nicholas Southwood and Lina Eriksson, 'Norms and Conventions', *Philosophical Explorations* 14/2 (2011), 195–217; Geoffrey Brennan, Lina Eriksson, Robert E. Goodin, and Nicholas Southwood, *Explaining Norms* (Oxford: Oxford University Press, 2013); Kai Spiekermann, 'Review of Explaining Norms', *Economics and Philosophy* 31/1 (2015), 174–81 and Laura Valentini, '"When in Rome, Do as the Romans Do": Respect, Positive Norms, and the Obligation to Obey the Law' (unpublished manuscript).

(We shall often refer to them as 'pre-contractual' rights and duties.) But if our most fundamental rights are *natural* in that they are independent of any social framework, in what sense could social rules affect their distribution? The answer that we develop here appeals to our power (i.e. our second-order right) to undertake duties towards others and exempt others from their duties towards us, by a voluntary (or at least non-coercive) habitual obedience to social rules.[4]

Social rules matter for the distribution of rights in two types of cases. First, consider natural rights whose scope is undetermined. Although some rights, like the right not to be killed, have a more or less definite scope even in the state of nature, other rights, like the right to private property, do not.[5] Their precise scope cannot be determined in the absence of an institutional structure.[6] Second, even when the scope of natural rights is definite and determined, social rules might affect their scope and distribution. In both types of case, by habitually following social rules we exercise our powers to undertake new duties and to exempt others from their 'old' ones.

Consider first the role of social rules in cases of indeterminacy. Let us assume, following Locke and others, that some property rights are pre-contractual. In particular, the right to valuable objects that were produced through one's labour exists independently of any social framework. However, suppose that once we move away from this core, we soon enter into areas of indeterminacy. Who owns an unidentifiable lost article—the original owner or the finder? Advocates of natural property rights might believe that if an owner loses an unidentifiable

---

[4] Let us comment about *other* social mechanisms that determine a distribution of moral rights and duties where voluntary, habitual obedience plays *no* role. Social rules might (first) harmonize our interests through resolving coordination problems; they might (second) instantiate abstract values; and they might (third) determine one way out of many things to do together. In more detail, conventions provide moral reasons because, as David Lewis noted in *Convention* (Cambridge, MA: Harvard University Press, 1969), they solve coordination problems, thereby promoting the well-being of all relevant parties. Moreover, as Andrei Marmor argued in *Social Conventions* (Princeton: Princeton University Press, 2009), conventions provide concrete ways of instantiating abstract values, such as respect. To be respectful of each other is to be polite; yet the rules of etiquette are thoroughly conventional. Our reasons for following them are the reasons we have to be respectful. Third, sometimes 'there is a need for some principle to govern a particular kind of activity, but there are a number of different principles that would do this in a way that no one could reasonably reject'; hence, when one of these principles is accepted in a given community, 'it is wrong to violate it simply because this suits one's convenience' (Scanlon, *What We Owe to Each Other*, 339). These social rules shape the moral landscape without interacting with our second-order rights and, as such, they do not interest us here. (We borrow this discussion from Tom Dougherty, 'Moral Indeterminacy, Normative Powers and Convention', *Ratio* 29/4 (2016), 448–65).

[5] There are views of natural rights that distinguish between the right to life and the right to private property (which has no definite meaning in the state of nature). See Arthur Ripstein, *Force and Freedom: Kant's Legal and Political Philosophy* (Cambridge, MA: Harvard University Press, 2009), 168.

[6] Another case of an indeterminacy that social practice removes arises from the phenomenon of imperfect duties. Once imperfect duties are institutionalized, certain positive special obligations are established, to which certain positive rights correspond. For example, one aspect of institutionalizing an obligation to care for children under particular social circumstances might be to assign social workers a positive obligation to monitor specific children at risk. See, Onora O'Neill, 'Children's Rights and Children's Lives', *Ethics* 98 (1988), 448–9 and 458.

article she thereby loses her right over it, merely because there is a social rule that whoever finds the lost article appropriates it. We suggest, in other words, that the moral rights and entitlements in the case of lost items are pre-contractually indeterminate, and that they are determined by social rules. These rules usually state that when a lost article has no identifying marks, the owner loses her right to ownership; whereas if the article bears such marks, she does not. (Yet even in the latter case, the owner's right to the item is not unconditional. If she does not seek to materialize it for a long time, her right is lost.)[7]

The logic behind this arrangement is that *ex ante* it is good for everyone. A prohibition against using lost articles that have no identifying marks would be a waste of resources; the owner would not be able to identify her item and reuse it, while all others would be barred from using it too. A rule according to which lost articles are disowned is fair because habitual obedience to it does not generate or sustain unfair inequalities. In fact, at least *ex ante*, the rule benefits all equally, irrespective of any arbitrary feature they might have (talent, social status, gender, age, etc.). The justification for the 'lost articles rule' might be put in Scanlon's terms: 'in considering whether a principle could reasonably be rejected we should consider the weightiness of the burdens it involves for those on whom they fall, and the importance of the benefits it offers for those who enjoy them leaving aside the likelihood of one's actually falling in either of those two classes'.[8]

But unlike Scanlon's reasoning, our version of contractarianism treats the mutual benefit condition (hereafter *Mutual Benefit*) as a symptom of free acceptance. A social arrangement is morally effective only if it enjoys a high level of habitual obedience. And if the rule is beneficial, it is presumably true that habitual obedience to it manifests free acceptance of the rule. Hence, the arrangement the rule dictates is morally effective; it determines the moral rights and duties vis-à-vis the issue it regulates. We define these conditions more accurately in what follows.

Let us turn to a second type of case in which social rules determine the distribution of moral rights. Free acceptance of a rule might determine a distribution of rights, even where in its absence, there is no indeterminacy in their scope. Consider:

ACCIDENT 1: Because of an unexpected threat, driver A must violate a basic traffic rule and move to the wrong side of the road, otherwise A will get herself killed. However, if A does so, she will hit and kill driver B who is following the rules and driving on the right side of the road.[9]

---

[7] For a more detailed discussion, see Benbaji, 'A Defense of the Traditional War-Convention' (2008).
[8] Scanlon, *What We Owe to Each Other*, 208.
[9] See the discussion of these examples in Yitzhak Benbaji, 'Justice in Asymmetric Wars: A Contractarian Analysis', *Law and Ethics of Human Rights* 6 (2013), 157–83.

In this case, most people would agree that A is morally prohibited from moving to the opposite lane, even if she must pay with her life to remain in her lane. Now consider the matter from the point of view of the other driver:

ACCIDENT 2: he same scenario as in ACCIDENT 1, but out of fear of the threat posed to her, A decides to move to the other lane. This time, if B hits A's car, it is A who will die. B can prevent the crash by moving to the other side of the road, but she would then fall into an abyss and get herself killed.

B is under no obligation to move aside and risk her life. She may keep to her lane even though she realizes that she will hit A's car and kill her.

From a pre-contractual moral point of view, A and B are equally situated. They are both engaging in permissible activity (driving) and, insofar as they put the other in peril, it is on account of factors beyond their control. Nevertheless, in both ACCIDENT 1 and ACCIDENT 2, there is moral asymmetry between the two drivers, which is determined by completely conventional traffic laws. To appreciate the crucial role of such laws in these cases, contrast them with the following lawless state of affairs:

ACCIDENT 3: A is driving her jeep in the Sahara desert. There are no marked paths or routes and no accepted rules about right of way etc. Because of an unexpected obstacle—an animal standing on the path—driver A must move to the left side of the path, otherwise she will get herself killed. However, if A does so, she will hit and kill driver B who is approaching from the other direction.

Here, it seems that A is under no obligation to stick to 'her' lane in order to prevent B's death. Of course, the same would apply to B, if she faced a similar threat. This means that A and B are symmetrically situated vis-à-vis each other. We tend to believe that in these cases, each has a right—a liberty right—to act in a way that would save his or her life, definitely by merely moving aside, but probably even by directly attacking the other driver to avoid the risk of a lethal crash.[10] Others might offer other solutions that respect this symmetry, like fair procedure.[11] The difference between ACCIDENT 3 and the former cases is

---

[10] See Chapter 1, fn. 44, for a further references. Let us note that if the drivers in this scenario could communicate, it would of course be better (it would actually be morally *required*) that they seek a fair way of reducing the expected harm—one that would guarantee each an equal chance of survival. But such communication is not possible in the case we are imagining.

[11] For further discussion, see Victor Tadros, *The Ends of Harm* (Oxford: Oxford University Press, 2011), 197–216; Gerhard Øverland, 'Contractual Killing', *Ethics* 115 (2005), 692–720; Susanne Burri, 'The Toss-Up Between a Profiting, Innocent Threat and His Victim', *Journal of Political Philosophy* 23 (2015), 146–65; and Helen Frowe, 'Equating Innocent Threats and Bystanders', *Journal of Applied Philosophy* 25 (2008), 277–90.

that in ACCIDENT 3 neither driver could be said to be driving in 'her' lane. In the absence of social rules, no one has a moral claim on any particular lane as being necessary for his/her survival. Therefore, neither of the drivers can be said to have lost her right to life by driving in the lane of the other.

Note that this verdict would change if *both* drivers were to die. In the Sahara circumstances, A has a natural right to prefer her own life over that of B, but not to act in a way that would uselessly kill both. If the only feasible option before A is to get herself killed by crashing into a standing animal or getting herself *and* B killed by crashing into B, she has no right to move into B's lane.

However, the fact that social rules make a moral difference, like in ACCIDENT 1 and 2, does not mean that they make *all* the difference. Consider:

ACCIDENT 4: Like ACCIDENT 1, but this time the reason that A has to move to the left lane is that B intentionally blew up the hillside and caused the right lane to disappear in the landslide.

In this scenario, the driving convention seems irrelevant to the distribution of rights. A has a right to drive on the left even if this involves hitting B. The crucial point is then: if social rules apply, they generate moral asymmetry even if the lives of the drivers are at stake.

In response to the ACCIDENT cases, one might argue that of course traffic laws—a paradigm of social rules—make a moral difference, but this is simply because there is a moral obligation to obey the law. Whoever accepts the existence of this obligation is committed to the view that violation of the law is at once a legal offence and a moral wrong, and as such ACCIDENT 1 and 2 merely flesh out some consequences of such violation. *Social Distribution*, therefore, is just a corollary of the moral obligation to obey the law and offers no interesting alternative to *Individualism*.

This argument, however, is a gross misunderstanding of the proposal we are making. To see why, consider again ACCIDENT 1. If A's duty to stick to the right lane is just an instance of her general obligation to obey the law, then we'd have to describe the moral dilemma at hand in the following manner: 'On the one hand, driver A has a natural liberty-right to move to the other lane, just as in ACCIDENT 3, even if that would lead to the killing of driver B. On the other hand, she has a moral duty to stick to her lane because that is the law and there is a moral obligation to obey the law.' But this would mean that the moral wrong that A would commit if she chose to move to the opposite lane would be pretty minor indeed. Most of us violate the traffic laws on a daily basis; usually, nobody considers such violations a serious moral issue. If this were the dilemma A faced, then if she moved to the opposite lane and killed B, her moral transgression would not consist of that killing, but rather would be limited to having violated the law. In contrast, we posit that traffic laws—as well as other social rules—determine (under specified conditions) *what we morally owe to each other.*[12] Given the way the practice of driving is shaped by the traffic laws, if A moved to the opposite lane, she would be committing a moral analogy to homicide. Her moral dilemma, then, is not between preserving her life and violating the moral obligation to obey the law, but between preserving her life and *wrongfully killing a person*; viz., violating his right to life.

2.3 The Conditions for the Moral Effectiveness of Social Rules

Our analysis of the ACCIDENT cases and of the social practice concerning lost items was intended to provide initial motivation for *Social Distribution*—the view that social rules partly determine the distribution of rights and duties; the ACCIDENT scenarios show that this view applies to fundamental pre-contractual rights whose scope is determined. But surely not all social rules have such a moral effect (e.g. the social rules defining the practice of slavery cannot change the moral rights held by slaves). In the present section, we explain why some social rules are morally effective while others are not. We elaborate on three conditions for such effectiveness, which we have already had the opportunity to introduce: *Mutual Benefit, Fairness,* and *Actuality.*[13]

We start with the condition of *Mutual Benefit*. As we indicated, traffic laws are mutually beneficial to all participants in this social practice, although in the cases discussed above some parties are doomed to lose because of them. Driver A in ACCIDENT1, for instance, is going to die. How could anyone suggest that the relevant traffic laws are beneficial to *her*? In a slogan that must be detailed and made more rigorous, the answer is that the test of *Mutual Benefit* is not an *ex post* condition but an *ex ante* condition. The right question to ask is not whether the social rules benefit all parties in all possible circumstances, but whether, given the evidence available when the parties agree on these rules, the expected benefit to them is higher than other feasible arrangements (or, at least, high enough—as we explain below). The question to be asked about participants is what they would rationally conceive as mutually beneficial *before* they get themselves in all kinds of trouble like the scenarios described in the ACCIDENT cases. What benefits them *after* a very improbable event, once they found themselves in an unfortunate situation, might be irrelevant.

Similarly, a rule that fixes the speed limit at 70 miles per hour on highways is mutually beneficial to all participants in this practice even though, had the speed

---

[12] On the idea that traffic laws usually constitute moral duties, see also George I. Mavrodes, 'Conventions and the Morality of War', *Philosophy and Public Affairs* 4/2 (1975), 117–31: 'it seems likely that different laws would have generated different moral duties, e.g. driving on the left' (126).
[13] For a discussion of similar conditions, see Yitzhak Benbaji, 'The Moral Power of Soldiers to Undertake the Duty of Obedience', *Ethics* 122 (2011), 43–73.

limit been lower, some would not have been victims of car accidents. The 70 mph rule has greater expected benefits for all relevant parties than any other alternative—which means that the victims of these accidents are *ex ante* beneficiaries of the high-speed-limit regime, despite the fact that a low-speed-limit regime would have been in their *actual* interest.

At first glance, the *ex ante* perspective looks as if it appeals to the idea of hypothetical consent. This is because to ask what was beneficial at some imagined point of time distinct from the actual point is to ask what rational drivers would agree to under such circumstances. However, the *ex ante* perspective to which we appeal here is not committed to anything like Rawls's veil of ignorance. When parties to a contract take the *ex ante* position, they do not abstract themselves from their individual features (gender, social-economic status, religion, and so on), like in Rawls's Original Position. The kind of uncertainty that is built into the *ex ante* perspective is the uncertainty actual people share about the future. This uncertainty constitutes the prudential reason to enter into contracts. In assessing the benefit of the contract, the parties maximize expected benefit (the sum of the value they attach to possible outcomes of entering a contract multiplied by the probability that these outcomes will transpire) rather than the value of the actual outcome of entering the contract. Although the parties have some knowledge about what might happen to them, they are unable to determine anything more than expected benefit.

The status of an arrangement such that all relevant parties (all drivers, for instance, or all citizens) would accept it *ex ante* is a major reason for regarding the outlook we propose here as contractarian. Yet the ultimate basis for the legitimacy of a social arrangement and for the distribution of rights and duties it entails lies not in its contribution to overall utility (however defined) but in its free acceptance within society. Presumably, those affected by the arrangement actually accept it *because* they rationally expect to benefit from it. That is, the habitual obedience to social rules within society manifests free (tacit) consent to the regime these rules set.

Most social practices do not benefit all members equally and cannot reasonably be expected to do so *ex ante*. Equal benefit is not what we have in mind here. Rather, roughly, a set of social rules is mutually beneficial if the state of affairs where the rules are followed is Pareto superior to a state of affairs in which the relevant parties tried to follow pre-contractual morality. For social practices to pass this test, they must make at least some parties better off, compared to circumstances in which the parties tried to follow pre-contractual morality.

Note that as the traffic laws example suggests, the set of rules governing some particular field of activity is not necessarily the only possible one, or even the best. It might turn out that, for some physiological reason, a convention of driving on the left side of the road is safer than a convention of driving on the right. Nonetheless, the latter is sufficiently good; therefore, if it is actually followed in a given society, it is morally effective. The arrangement need not benefit the relevant parties in the *best* possible manner, only in a reasonable and satisfactory way.[14]

Saying that some arrangement is effective although it is not the best possible one creates a troubling feeling of compromise. In a sense, this feeling is warranted. After all, there is a better social arrangement that would—had the world been different—contribute more to the wellbeing of the relevant parties and to the protection of their rights. But the world is *not* different; hence, in the actual world, affirming this less-than-ideal arrangement is often the right thing to do.[15] Moreover, in many cases, major revisions in social arrangements are predicted to end up causing more harm than benefit, with the weakening of these arrangements instead of their improvement. Thus, one might say (somewhat paradoxically) that the actual code becomes the ideal one, partly because it is the one used and accepted in some cases.

So far, we have introduced no constraints on the kind of interests that the parties in the *ex ante* condition are allowed to take into consideration, no constraints on the kind of benefit that would satisfy the mutual-benefit condition. Surely, however, there are such constraints. As numerous writers have argued, some preferences—famously named by Dworkin's 'external'[16]—ought to be discounted when we shape social arrangements in light of utilitarian considerations. This means that the condition of *Mutual Benefit* would not be met if the benefit it secured to some parties depended on, say, the chauvinistic or racist preferences that they held. Therefore, by *Mutual Benefit* we mean only benefit unrelated to external preferences. This is a different way of saying that the *ex ante* perspective is that of rationally partial, yet decent, parties; rationally partial in caring about their own benefit, and decent in having basic respect for other people.

This point paves the way for us to present the second condition for the moral effectiveness of social rules; namely, *Fairness*. Some social rules distribute benefits and burdens unfairly, while others institute or support unfair or disrespectful social relationships. Imagine the following contract between slaves and masters: the slaves subject themselves to the duty to comply with their masters' orders, while, in return, the masters undertake the duty to refrain from violent coercion.[17] Both sides benefit from the agreement. The masters save the costs of using force, while the slaves are spared the cost of suffering. Given the actual circumstances under which the slaves and masters operate, accepting the arrangement would

---

[14] For the idea that choices might be justified even if they are only good enough and not the best, see, e.g., Michael Stocker, *Plural and Conflicting Values* (Oxford: Clarendon Press, 1990).

[15] Which is consistent with a cautions attempt to reform it, if one deems such reform necessary/possible. See Mavrodes, 'Conventions and the Morality of War', 127: 'One might simultaneously have a moral obligation to conform to a certain convention and also a moral obligation to replace that convention.'

[16] See Ronald Dworkin, *Taking Rights Seriously* (London: Duckworth, 1977), 234–8, 275–7.

[17] See David Gauthier, *Morals by Agreement* (Oxford: Clarendon Press, 1986), Chapter 7.

bring about a state of affairs that all parties prefer. Nonetheless, such a contract would not be binding because the background circumstances that it assumes are grossly unjust. Therefore, the expectation that the slaves respect the agreement has no moral bite. If the slaves can use force to drive their masters from power, morally they may do so.

To generalize, then, the fact that some social arrangement brings about a Pareto-improved state of affairs leaves open the possibility that the distribution of the relevant benefits among the parties is unfair. *Fairness* stipulates that an arrangement that creates or solidifies unjust inequalities in the distribution of a morally relevant good is invalid, and the contract in which it is embedded is not binding. The unfairness that invalidates a social contract is usually not merely the unequal distribution that benign free competition might generate. Rather, it requires an unequal distribution that is a matter of brute luck or inequality that results from, expresses, and advances unjust relations like exploitation, oppression, subordination, and so on.[18]

In order to identify unfair social arrangements that are nevertheless Pareto superior to the relevant baseline, we must answer three questions. (1) What is the relevant *unfairness-free* baseline? That is, what is the state of affairs that would be brought about if decent and partial parties tried to implement pre-contractual morality? Such clarity should help us to answer the next two crucial questions: (2) does the arrangement under assessment *maintain* the unfair inequalities in the actual state of affairs? And (3) does the arrangement *create* unfair inequalities of its own? Such questions are of special importance to our upcoming discussions of the prohibition on the use of force, the right to national defence, and the rules that govern wars between parties enjoying asymmetrical power. (As we noted, it is nearly axiomatic that the lost article arrangement is fair. This is because even if the initial distribution of goods is unfair, the lost article arrangement neither maintains the unfairness of the initial distribution nor is expected to create new unfair relations.)

We turn now to the third condition, *Actuality*.[19] In a sense, we get *Actuality* 'for free'—we follow Hart, who characterizes social rules as rules that receive habitual obedience and implicit acceptance (alongside other definitive features, of course). The rules that interest us here are morally effective, in the sense that *Social Distribution* alludes to—only if they are actually adhered to in a given community.[20] To be morally effective, they must actually be followed by most members of

---

[18] Some, most notably Elizabeth Anderson, argue that in and of itself, 'brute luck' does not generate injustice. See Elizabeth Anderson, 'What is the Point of Equality?' *Ethics* 109/2 (1999), 287–337.

[19] This aspect of our theory is indebted to David Lewis. In particular, based on his analysis, we assume intimate relations between conventions, rules, and norms. Lewis noted in *Convention*, pp. 88–100.

[20] See Mavrodes, 'Conventions and the Morality of War': 'In cases of convention-dependent obligations the question of what convention is actually in force is one of considerable moral import' (128).

the relevant community, be they drivers, finders of lost items, or combatants; the latter comprise our main focus in later chapters.[21]

But *Actuality* should be highlighted because of its role in explaining the fact that social rules are morally effective. Indeed, this condition can be defended in various ways. First, the fact that a set of rules is actually followed in some field creates expectations that all players who enter the field have a *pro tanto* reason to respect. The moral force of these expectations stems from their inherent justifiability, since the rules are mutually beneficial and fair. Once a practice meets these conditions, it creates legitimate expectations that the relevant parties have a (*pro tanto*) duty not to frustrate. These expectations constitute another reason to regard the view we are defending as contractarian. Conjoined to the conditions of *Fairness* and *Mutual Benefit*, *Actuality* entails that the expectations that the rules to be respected are legitimate.[22]

Second, *Actuality* is necessary because in many fields there is more than one reasonable set of social rules. The fact that of several possible sets of rules, only one is actually adhered to in a given community is what grants it the power to determine the moral rights of community members. The accepted arrangement is effective even if it is not the best one, provided that it exceeds a certain threshold of reasonableness. (Again, one might worry that *Actuality* makes social norms too contingent—shaped, as it were, by what the community happens to accept in practice rather than deduced from some eternal principles. We see no cause for concern here. To the contrary: this contingency confirms the 'down to earthness' of the social norms and the high chances that they will be respected.)

All of the above reasoning is important in understanding the bite of *Actuality*. However, the role that *Actuality* plays in our contractarian framework is much more principled: within this framework, the fact that mutually beneficial rules receive habitual obedience makes it presumably true that they are freely accepted. By freely accepting fair and mutually beneficial social rules, members of the society in question lose some of their natural rights. In other words, when people habitually follow these rules, they waive the relevant rights in exchange for expected benefits, under conditions of fairness. Our version of contractarianism goes a step further. By merely belonging to a society and participating in the social practices within it, individuals vindicate the presumption that they freely accept the mutually beneficial and fair social rules that underlie these practices. Hence,

---

[21] As argued by Matthew N. Smith, 'Terrorism as Ethical Singularity', *Public Affairs Quarterly* 24 (2010), 233, with respect to the laws of war, 'it is important to recognize the contingency of these norms of war, and in particular their dependence upon practice'.

[22] Indeed, 'social conventions' might designate coordinated behavioural regularities that lack normativity; yet once members expect others to conform to them, normativity emerges. Legitimate expectations generate *pro tanto* reasons to respect them. Note also that as Lewis clarifies (*Convention*, 63), *Actuality* typically implies that the conventions are known in a 'poor' sense; members of the relevant society know them only tacitly and are often unable to articulate them.

qua members of society, individuals are subject to the mutually beneficial and fair social rules that underlie the various social practices within this society.

To illustrate the relation between tacit acceptance (manifested in one's behaviour) and one's waiver of basic rights, consider the moral permission that boxers obtain to hit and often injure each other, in apparent violation of the adversary's natural rights. The permission is based on their voluntary relinquishment of (some of) these natural rights when they enter the ring. In most social contexts, such relinquishment is less conscious and less explicit; but it can, nonetheless, be ascribed to the relevant players. Similarly, by driving, or even getting into a car, a person joins a given social practice and undertakes to abide by the rules that govern it. These rules are *ex ante* mutually beneficial. Therefore, it can be presumed that she undertook the duty to follow them and allowed others to put her at risk, as long as they respect the driving convention. The fact that she would have agreed to a set of rules that implies losing a set of natural rights, coupled with the fact that she actually takes part in a practice governed by these rules, implies a loss of the rights in question.

The significance of being part of a social practice can be seen by looking at cases of innocent mistakes about its exact rules. The fact that some driver is mistaken about who has the right of way when driving on a narrow bridge is irrelevant to determining her legal (and hence her moral) right vis-à-vis other drivers. Indeed, these rules are the source of rights and duties in respect to others who are driving toward you in the opposite lane, as in ACCIDENT-like scenarios.

In a similar vein, consider a legal system that permits unilateral exit from marital relationships. Suppose that a religious couple gets married in this social context, but, at the time of marriage, both individuals hold the erroneous belief that their marriage creates an indissoluble relationship. Suppose they believe that the possibility of unilateral or even bilateral exit from a marital relationship is incompatible with the meaning of marriage. This belief is irrelevant to the moral status of their marriage. In fact, they do have a right to exit the marriage under the conditions specified by the law. To determine what moral liberties, claims, and duties are created by entering marital relations, one need not look into the heads of the participants engaged in this practice; the meaning of the rules defining this relationship is not to be found there. Once individuals consent to be married, they consent to the legal norms that define this institution. By this act of consent, they allow for the redistribution of the moral claims that they hold against each other—assuming of course that the norms in question are fair and mutually beneficial.[23]

---

[23] Some might infer from our position that consent is unnecessary and that it is sufficient that the role-holder willingly gets his paycheck. As John Simmons puts this view: 'One might have an obligation of fairness to do one's part within a cooperative scheme from which one has willingly benefited, where "doing one's part" consists precisely in performing the tasks attached to an institutionally defined role.' This can be true even where 'we have not strictly "accepted" those roles at all, but have only freely accepted the benefits provided by the schemes within which those roles are defined' (A. John Simmons,

Now, most philosophers would agree that rights might be waived even if people never explicitly agreed to the rules whose acceptance has this effect. They might argue, however, that the acceptance that people manifest through habitual obedience to the rules is too weak. Habitual obedience cannot generate a redistribution of rights. Tom Dougherty argues (with respect to indeterminacy cases) that 'by implicitly endorsing the code, citizens can impose on themselves determinate obligations to conform to the code, and withdraw any indeterminate claims that they have against someone who follows the code'.[24] We take a stronger position: the social norm does not have to be endorsed for it to reshape the distribution of moral rights and duties in the society in question. Such redistribution can stem from the mere act of non-coercive habitual obedience to the rule. Acceptance of the rule is very close to the knowledge that people 'around here' follow this rule, and that one is a member of the community in question. The contractarian view that we offer here implies that, in some cases, a person loses her rights just by being an active member in the society. If an arrangement is good and fair, one may presume that *all* members of the society in question accept it, even if they are reluctant and somewhat resentful of the social rule in question.

Thus, contractarianism loads the Hartian concept of acceptance with normative significance. It offers a set of conditions under which the acceptance of rules is presumably free. Presumably, those who subject themselves to a system of social rules freely accept it, if and only if it is mutually beneficial and fair. Contractarianism asserts that by freely accepting a legal system, individuals implicitly consent to be governed by it and waive some of their pre-contractual rights. Along the way, we relaxed the constraints on what it is like to freely accept a legal system: people who accept a social rule do not have to be conscious of all its details and definitely not to explicitly endorse it. They accept the rule by virtue of having a standing disposition to follow it qua members of the society to which they belong.

In this section, we sought to clarify and substantiate the claim that social rules affect the distribution of moral rights and duties. This is particularly palpable in cases in which social rules define social *roles*. This will be the topic of the next section.

2.4 Social Roles and the Moral Division of Labour

The rules that determine the relations between drivers apply equally to all of them. But some individuals—police officers, for instance—have a special status in

---

'External Justifications and Institutional Roles', *Journal of Philosophy* 93 (1996), 29). *Pace* Simmons, we take the view that by freely accepting the benefit, one makes it presumptively true that one accepted the terms of one's role.

[24] Dougherty, 'Moral Indeterminacy, Normative Powers and Convention', 250.

regulating the practice of driving. Likewise in other fields. Some people—teachers, bankers, policemen, etc.—are assigned legal or professional duties, privileges, powers, and immunities not assigned to others. In this section, we argue that these roles are central in determining, or in distributing, moral rights and duties.

Let's start with the following case:

DEAN: As the dean of a small yet excellent law school, it is my job to feed the grades that the students got for their final papers into the university computer. I am aware of the fact that not all professors invest the required time and effort in marking the papers and, consequently, there is a non-negligible chance that some students will have received a lower grade than they deserve. Assume that I can revise the grades given by the professors without this ever being disclosed.

Under these circumstances, do I have an obligation to review the papers of all students prior to feeding their grades into the university computer? Surely not. I have a right (a liberty-right) to suspend judgment about the quality of the papers and just feed the grades into the computer, or ask my assistant to do so. But what if I did look at some paper and was convinced that the grade was unfair? Assume that the relevant professor would never agree to change the student's grade and that I am the only person who could prevent this injustice from occurring. Nonetheless, we propose, I have a right not to change the grade. If I said to the student or to some colleague, 'I'm sorry, but this is just not within my authority', it would be a perfectly good answer. We offer the same response to cases where the stakes are higher:

POLICEMAN: A policeman receives an order from his superior to go out and arrest a person who is suspected of having committed some crime. The policeman happens to know the person and also to know about the crime, and is almost positive that this is the wrong suspect.

As a descriptive matter, the policeman is under a legal (and professional) duty to submit to the order. Many people, we surmise, would think that typically in a decent society, the policeman also has a moral duty to obey this order and carry out what he suspects is an unjust arrest. Others would disagree. But almost everybody would concede that he has at least a moral *right* to do so, which amounts to a right to act in a way that seems to violate the rights of another person. In the spirit of *Social Distribution*, the role of a policeman as defined by a long list of rules and as actually practiced in modern society changes the distribution of moral rights. Qua policeman, a person has a moral right to do to others what he would be barred from doing to them qua a regular human being.

Finally, consider a case where the stakes are really high, a matter of life and death:

UNJUST EXECUTIONER: An executioner has good reason to believe that the prisoner he is about to execute is innocent of the crime for which he has been sentenced to death.[25]

If the executioner does his job and kills the prisoner, he will probably be intentionally killing the innocent. Nonetheless, it seems that the prisoner does not have a claim-right against the executioner that he refrains from killing him, just like the arrestee does not have such a claim against the policeman coming to arrest him. Neither the policeman nor the executioner is seen as violating the rights of his respective victims.

In these examples, the agents know, or at least have good reasons to suspect, that the rights of the arrestees, students, etc. are violated; yet they retain the right—once again, the moral right—to act within their social roles. In the more typical cases, agents fulfilling social roles lack such knowledge; hence, they do not confront a clear conflict between the demands of their role and the requirements of justice. Still, their social role makes a significant difference: it grants them exemption from the standard moral demand to make sure, prior to a seeming violation of rights, that their action is morally justified. The dean is not expected to check the students' papers on his own; the policeman has no obligation to verify the evidence substantiating the arrest warrant; and the executioner is not expected to check for himself the evidence against the prisoners on the death row.

What emerges from these examples is a division of labour among various role-holders in society. The policeman is exempted from the duty to make sure that arresting the suspect is justified, since others in the society that institutionalizes this role are under a professional duty to do so.[26] Thus, according to the contractarian framework offered here:

Moral Division: If a system of social rules encompassing defined social roles is mutually beneficial and fair, the moral rights that members in this society possess and the duties to which they are subject are affected by this system. In particular, role-holders might have a moral right to follow the package of social rules that define their role, and (qua holders of such roles) to disregard moral reasons that pertain to their actions.

---

[25] We do not take a side in the debate about whether executions can be ever justified. We assume, though, that societies are not *indecent* by mere virtue of imposing capital punishment on occasion.

[26] The connection between professional duties and the division of labour is central in Arthur I. Applbaum, *Ethics for Adversaries: The Morality of Roles in Public and Professional Life* (Princeton, NJ: Princeton University Press, 1999). In Applbaum's understanding, underlying this division is the claim 'that some good ends are best produced under a form of social organization in which differentiated actors pursue more narrow aims, rather than aiming directly at the good end that is the purpose of the institution' (197–8).

To defend *Moral Division,* we appeal to Raz's analysis of legitimate authority. Typically, in decent societies, the rules that define a person's social role manifest a legitimate authority over her. That is, generally, they constitute a second-order reason for her to fulfil her professional duties *without* making sure that what she ought to do (as a matter of professional duty) is morally justified. In cases where the division of labour is authoritative, a policeman has a reason not to be guided by a cluster of first-order reasons for or against arresting a suspect, since his commander is typically better positioned to respond to these reasons. More generally, the directives issued by a legitimate authority pre-empt the need for further practical reflection by individuals who are subject to this authority. This is usually because individuals have only limited access to the first-order reasons that apply to them. Given that the relevant authorities have much better access to these reasons, following their instructions is a much safer strategy than engaging in deliberation on the merits of the case. This is a 'service conception' of authority:[27] the legal system in decent states usually serves its addressees by mediating the first-order reasons provided by pre-contractual morality.[28]

Now, our account stresses that to be morally effective, a division of labour ought to be fair and mutually beneficial; that is, it should *usually* serve its addressees. However, crucially, it does not appeal to the service conception of authority in order to explain the moral efficacy of the social division of labour. This is because the reach of the service conception of authority is too limited. Take legal systems. They are robust, and cover a huge variety of circumstances. Powerful interpreters extend them to changing circumstances in one way or another. Therefore, it is hard to believe that in all cases that the law covers, it has legitimate authority over most of its addressees. True, the rules that command the obedience

---

[27] See Joseph Raz, *The Morality of Freedom* (New York: Oxford University Press, 1986), 39. For an illuminating discussion of Raz's views, see Scott Shapiro, 'Authority', in Scott Shapiro and Julius Coleman (eds.), *The Oxford Handbook of Jurisprudence and Philosophy of Law* (Oxford: Oxford University Press, 2004), 382–440. Raz's conception of the normal justification of rules has been criticized from various perspectives. See a summary in Shapiro, Section IV, which includes a helpful presentation of Michael S. Moore, 'Authority, Law and Razian Reasons', *Southern California Law Review* 62 (1989), 866–7. See also Chaim Gans, 'Mandatory Rules and Exclusionary Reasons', *Philosophia* 15 (1986), 373–96, William A. Edmundson, 'Rethinking Exclusionary Reasons', *Law and Philosophy* 12 (1993), 329–43 and Merten Reglitz, 'Political Legitimacy Without a (Claim-) Right to Rule', *Res Publica* 21/3 (2015), 298–9. Recently, Laura Valentini expresses this criticism as follows: 'the existence of a positive norm, namely of a widely accepted "ought" does not explain why one has a pro tanto obligation to perform a given act.' She attacks 'the hypothesis that we have obligations to *obey* positive norms: i.e., to perform or avoid an action *because* there are positive norms prescribing or prohibiting it.' Rather, 'we have contingent reasons to *act in line with* the norms' prescriptions... the obligation to obey positive norms must be "content-independent"' (Valentini, '"When in Rome, Do as the Romans Do" Respect, Positive Norms, and the Obligation to Obey the Law'). The same point was made forcefully by Victor Tadros, who argues that subordinates 'are not required to follow orders; at least if by "following orders" we imply, as we surely do, that the order itself must figure in some way in the practical reasoning of the subordinate' (Victor Tadros, 'Anarchic War']unpublished manuscript[). A systematic defence of the service conception of authority lies beyond the scope of our discussion here.

[28] This is consistent with the possibility that some legal subsystems in a decent state have no legitimate authority.

of policemen usually bring about good results. Occasionally, however, policemen can know *ex ante* that the rules command them to do the wrong thing. And, we submit, even in these cases, policemen might have a moral right to follow orders. Indeed, they have the right to follow orders despite the fact that, in these rare cases, the rules have no (Razian) authority over them. Hence, the legitimate authority that these mutually beneficial and fair rules typically have is insufficient to explain the *scope* of the moral effect of these social rules and the social roles that they institute. Social arrangements are morally effective even where they do not serve their addressees.

Hence, we stress that *Moral Division* is contractarian. As Michael Hardimon observes, 'what one signs on for in signing on for a contractual social role is a package of [norms], fixed by the institution of which the role is part'.[29] And, according to contractarianism, it is by accepting social rules that members in society lose their moral pre-legal rights against, for instance, unjust arrest. Role-holders become subject to these norms by virtue of their acceptance and the acceptance of others of the norms that define their roles. The legitimate authority of the division of labour is important: it constitutes the basis for the presumption that the arrangement in question has been freely accepted by those who habitually subject themselves to it. Yet, the normative work is done by the tacit acceptance of the arrangement, which can be presumed to be free.

Moral Division applies to theories of distributive justice in a way that is particularly illuminating. Let us look at a central question regarding Rawls's theory of justice.[30] Famously, Rawls argues that the basic structure of society should be designed so that social and economic inequalities are to the greatest benefit of the least well-off members. In his opinion, the mere fact that some members of society have more than others is not troublesome, provided that the arrangement improves the situation of the least well off, in comparison to how they would do under any other distributive scheme. In particular, a tax regime ought to allow inequality in net income to the extent that the resulting incentives would tend to raise the lowest income. A tax regime is unjustified if it eliminates social inequalities by reducing the net income of talented people to that of the less talented. For, arguably, such a distributive scheme distorts the incentives of talented individuals: they would not work as hard and would not contribute as much to the social product.

Thus, Rawls's theory of justice implies that in the real world, agents operating in the market should have the legal right to exhibit acquisitive behaviour, if this supports the efficiency of market activity. In fact, for Rawls, what we owe to each

---

[29] Michael O. Hardimon, 'Role Obligations', *Journal of Philosophy* 91/7 (1994), 354. These norms are interpretive: people can reasonably argue about the proper interpretation or understanding of role terms' (p. 336).
[30] For a detailed discussion of this Rawlsian theme, see Gerald A. Cohen, 'Where the Action Is: On the Site of Distributive Justice', *Philosophy and Public Affairs* 26/1 (1997), 3–30.

other as fellow citizens through our common institutions is very different from what we owe each other as private individuals. As agents operating in the free market, we have a right to advance our interests and maximize our profits. In our capacity as citizens, we are under duty to establish and maintain social institutions whose aim, *inter alia*, is to promote social justice among members of society. For Rawls, this division of labour—the partial and self-interested behaviour of individuals operating in the economic realm, on the one hand, and the concern for justice on the part of citizens that struggle for the justice of social institutions, on the other—is essential to real-world just societies.[31]

G. A. Cohen finds this picture incoherent. In his view, it makes no sense to require citizens to pursue egalitarian justice through social institutions while exempting them from this pursuit in the private spheres of their lives. The realization of a more just society should guide people in their private lives too, in their capacity as private citizens; hence, people must lead much less acquisitive lifestyles and should transfer resources to the less advantaged, so as to reduce the effects of brute luck on the lives of fellow citizens.[32]

In the terms used earlier, Cohen denies *Social Distribution* and *Moral Division*. He claims that if justice requires a certain distribution of goods between people, then each individual has an obligation to advance it, regardless of the contingent social role that he or she plays. If you are the minister of finance or even a regular member of parliament, you can obviously do much more to promote justice in society than if you are an ordinary citizen. But that does not mean than as an ordinary, private citizen you are allowed to ignore such considerations altogether and care solely about your own self-interest. The battle for (distributive) justice is one in which we must all play our part.

However, consider:

BANKER: You justifiably believe that justice requires the imposition of heavier taxation on firms but, in the near future, these taxes are unlikely to be levied for good reasons. Public opinion goes against raising taxes, and therefore the costs of such a change are too high at the moment. Fortunately, you work in a bank that serves big firms as well as clients who barely survive. You could easily transfer some money from the big firms to the least well-off clients, thereby making your modest contribution to the realization of a more just world in Rawlsian terms.

Cohen's logic leads to the conclusion that you are under a moral obligation to do so, which is hard to accept. Rawls's theory of justice seems to offer a more plausible response to BANKER. If the system in general is mutually beneficial

---

[31] See Thomas Pogge, 'On the Site of Distributive Justice: Reflections on Cohen and Murphy', *Philosophy and Public Affairs* 29 (2000), 159.
[32] Cohen, 'Where the Action Is: On the Site of Distributive Justice', 26–8.

and fair (the distribution is unfair, but the tax regime does not solidify the unfairness), the banker is exempted from deliberating on whether or not such a transfer of funds would make the society more just, and has a right to stick to the rules of the bank with all the partiality that they imply.[33] The question of who has a moral obligation to promote distributive justice and who is allowed to ignore it is determined by the social role that an individual contingently fulfils.

2.5 The Conditions for the Moral Effectiveness of Social Roles

We now appeal to the conditions for moral efficacy presented above in Section 2.2, to argue for the possibility of the moral division of labour in a more systematic way.

A well-designed regime that institutes social roles through power-conferring rules is likely to bring about a Pareto-superior outcome to the baseline outcome (viz., an outcome in which most decent but partial individuals try to follow their pre-contractual duties and take advantage of their pre-contractual rights). This is especially so because a well-designed social regime would institute a division of labour that enables epistemic and moral limitations to be overcome. The optimal rules divide the labour of attending to moral and factual considerations among different actors in society, according to their epistemic and moral virtues and their means. The division is sensitive to the assumed expertise of role-holders in the relevant fields; hence, it is both natural and learned. If the system is well designed, the agents assigned to specific jobs are better equipped than most others to carry them out in terms of their professional knowledge, experience, and motivation. Judges, for instance, are in a much better position to decide on whether A unjustly harmed B than B is, than any of her friends are, and probably than most members of society are. Thus, when policemen have a right to disregard first-order reasons pertaining to an arrestee's assumed guilt, it is because they can reasonably believe that the court issuing the warrant is (a) basically just and (b) epistemically more reliable than they are in reaching the right verdict in the case at hand.[34]

Epistemic division of labour is not the only reason to act within the capacity of defined social roles within a well-designed regime. By dividing the labour, we

---

[33] *Moral Division* is also the key for solving the well-known problem of how lawyers are morally permitted to defend an accused criminal without being convinced that he is indeed innocent, or even if they strongly believe that he is *not*. For the development of this argument, see Frans Jacobs, 'Reasonable Partiality in Professional Ethics: The Moral Division of Labor', *Ethical Theory and Moral Practice* 8, (2005), 141–54.

[34] For a similar argument, see Marquez Xavier, 'An Epistemic Argument for Conservatism', *Res Publica* 22 (2016), 405–22. Xavier rightly emphasizes the conservative implications of the above division of labour. For discussions about the division of cognitive labour in other areas of philosophy, especially in the philosophy of science, see Philip Kitcher, 'The Division of Cognitive Labor', *Journal of Philosophy* 87/1 (1990), 5–22; Johanna Thoma, 'The Epistemic Division of Labor Revisited', *Philosophy of Science* 82/3 (2015), 454–72; and David Eck, 'Social Coordination in Scientific Communities', *Perspectives on Science* 24/6 (2016), 770–800.

overcome coordination problems. Policemen assigned to maintain order in some public event have the legal authority—which immediately translates into moral authority—to arrest hooligans. In contrast, ordinary citizens standing nearby lack such power even if they have the same knowledge that the policemen have and the same capacity to 'arrest' the hooligans. The division of labour is mutually beneficial because if ordinary citizens had such power, that would open the door to social instability and violence. Hence, except for under extreme circumstances, ordinary citizens are barred from 'arresting' hooligans even when such an arrest would be justified in terms of the protection of rights. At times, then, the reason for dividing labour—for assigning social tasks to specific agents or agencies—is that if the task were to be left to any and all, it would be carried out in a less effective way with potentially serious negative repercussions.

To be morally effective, the division of labour embedded in social arrangements should not maintain or create unjust inequalities between groups or between individuals. One implication of the fairness requirement is that when social institutions such as courts or municipalities systematically discriminate on the basis of race, religion, or gender, the power-conferring rules that empower them to do so are morally ineffective. Policemen have no moral right to follow the directives of judges in a corrupt system, without verifying that their verdicts are just. Members of a police force that routinely rounds up citizens of a perceived inferior race have no (moral) exemption from making sure that every arrest warrant they receive is morally justified. The same applies—all the more so—to executioners.

As we noted, social rules by which social roles are instituted are 'actual' by definition. The social role of a banker would not exist if banks did not play the crucial role that they do in modern economies; the role of a policeman would not exist without a functioning police force that carries out well-known tasks; and so on.

If the package of norms that defines roles is fair and mutually beneficial, the habitual obedience of role-holders to the rules that define it generates legitimate expectations that they behave in a certain way. Therefore, they have a *prima facie* obligation to live up to these expectations. And vice versa. Bankers and policemen have legitimate expectations from their clients, which, other things being equal, their clients ought to uphold. As we have already stressed, the role of habitual obedience and acceptance in our framework runs deeper. If the power-conferring rules are mutually beneficial and fair, their acceptance bases a presumption that the rules in question are *freely* accepted. The fact that the rules are freely accepted within society entails that, if they are fair, these rules affect the distribution of the rights and the duties of the role-holder and ordinary citizens: a citizen has no claim against a policeman not to arrest her before he first verifies that the arrest is morally warranted.[35]

---

[35] Of course, we realize the risk of role-holders abusing the special rights granted to them, but, as Nagel notes, the corrupting effects of such rights 'would not be so unless there were something to the

To sum up this point, the contractarian construal validates the moral division of labour in the following way. First, it observes that social arrangements are built on an institutional division of labour between various role-holders. Contractarianism then assumes that the fact that the arrangement is mutually beneficial grounds the presumption that the arrangement in question is universally and freely accepted. Individuals who are, in fact, freely governed by this legal system tacitly know that this system allows policemen to be engaged in presumptively wrong actions like arresting other people. The acceptance of the rules within society, which is manifested in the habitual obedience to the rules that define them, generates a moral right to undertake such actions if the rules that permit such actions are necessary to attain the goal of the system and if the system is fair and mutually advantageous.

All this means that role-holders have a right to ignore first-order reasons against actions that they are required to do within their role even if absent the rules that define the roles, such actions would violate the rights of some people. Presumptively, the victims of these supposedly wrong actions have accepted the relevant social system and, by doing so, waived their right not to be harmed by the relevant role-holders. That is why policemen violate no right that the arrestee has against them as long as they follow orders from a de facto authority (under the conditions mentioned above). Under normal circumstances, arrestees can be assumed to have waived their right to have the policemen verify the justifiability of their arrest before carrying it out.

What should *Individualism* say about cases like POLICEMAN? Would it say that policemen have a duty to verify for themselves that the suspicions against suspects are convincing before they carry out the arrest? Would it say that policemen must refuse to carry out arrests when they have doubts about their justification?

One might reply to these questions in the positive, arguing that the amount of scrutiny that should be devoted to assessing the justification of disobedience is proportional to the likelihood of mistakes in the relevant field and the seriousness of the injustice that would follow. For instance, according to a recent study, around four per cent of those sentenced to death in the United States are innocent of the crimes ascribed to them.[36] This means that the likelihood of being unjustly punished is non-negligible. Since such injustice is very serious, this proposal would entail that law-enforcing agents—policemen, prison guards, executioners—would have to make sure, prior to carrying out court decisions, that they are not accomplices

---

special status of action in a role' (Thomas Nagel, 'Ruthlessness in Public Life', *Mortal Questions* [New York: Cambridge University Press, 1979], 76).

[36] See Samuel R. Gross, Barbara O'Brien, Chen Hu, and Edward H. Kennedy, 'Rate of False Conviction of Criminal Defendants Who Are Sentenced to Death', *Proceedings of the National Academy of Sciences in the United States* 111/20 (2014), 7230–5.

in this injustice. Namely, these agents would have no right to disregard the first-order reasons pertaining to the cases of which they are in charge.

Under circumstances in which the rules have legitimate authority over role-holders, we should hope that role-holders would avoid taking this advice. This would weaken their ability to carry out their roles, and, most probably, will not yield better results. Role-holders should take into consideration not only the likelihood of mistakes made by the relevant authority, but the likelihood of themselves doing any better given their much poorer expertise, experience, and information. It is, then, in the interest of all relevant parties to fix the package of norms that defines the contractual social role of a police officer such that it includes a right to disregard some reasons that apply to her, if it is the duty of other role-holders to be guided by them.

Thus, individualists do not have to be anarchists. They need not deny the existence of legitimate authorities that serve their addressees. Individualists would probably agree that, as a rule, role-holders are not accountable for failing to rescue innocent people from imprisonment or execution. But what individualists offer such role-holders is merely an excuse (or justification in the evidence-relative sense), not a moral right (or justification in the fact-relative sense). According to *Individualism*, role-holders who inflict harm by arresting people or hitting them can never be said to have acted within their rights if the arrest and the harm was unjust.[37]

Contractarianism offers a different response. The notion of excuse does not fit because, under the circumstances discussed, the role-holders would not be wronging the victim. The executioner has a right (against the innocent victim) to harm her because the victim has waived the relevant right that she has vis-à-vis the executioner. If a person is innocent of the crime for which he was convicted, how could it *not* be morally wrong to take away his freedom and lock him up in a prison cell? The answer—to recap—is based on the premise that a social contract includes a trade of rights by all participants in exchange for the benefits they expect to gain.

In concurrence with *Individualism*, then, we cherish the value of human rights and regard the constraints they impose on behaviour as crucial. But, unlike *Individualism*, we believe that individuals can lose these rights vis-à-vis specific role-holders in specified ways when they live in an organized society.[38] To join a social practice *is* to trade rights in exchange for critical benefits. Once this trading in rights is morally effective, the duties incumbent on those fulfilling social roles change. While private citizens are under an obligation to make sure that their

---

[37] In the context of war, McMahan concedes that many soldiers fighting in unjust wars are fully excused, others culpable to varying degrees, and a small minority 'may even deserve punishment' ('The Ethics of Killing in War', 730).

[38] In more technical terms, individuals have the Hohfeldian power to release specific role-holders from various duties that the latter owe them.

actions do not violate the rights of any other person, the same individuals acting in the capacity of their social roles are exempted from this duty and instead are expected to do their job; to arrest people against whom courts have issued warrants, to invest the money of rich clients with the aim of maximizing gains, and so on.

2.6 A Moral Duty to Fulfil Professional Duties?

So far we have developed a version of contractarianism according to which its central proposition—*Moral Division*—entails a (liberty-)*right* to act in ways contrary to the pre-social distribution of natural rights, not an *obligation* to do so. However, one might claim that this version is too modest. If society depends on a division of labour among its members, then maybe it would be agreed *ex ante* that role-holders—or at least some of them—would have an obligation to stick to their professional roles, not merely a right to do so.

To start unpacking this question, we note an ambivalence regarding the nature of that which role-holders have a right—or an obligation—to disregard. It might be the burden of *investigating* into the relevant first-order reasons, and it might be the burden of *acting* in accordance with what they believe to be the right thing from some morally impartial perspective.[39] Consider again prison guards. The exemption granted to them might be interpreted as a permission to lock an inmate in his cell without first verifying his blameworthiness, or as a permission to do so even if they believe that the inmate is actually innocent of the crimes ascribed to him. When we ask whether *Moral Division* entails a right or an obligation, we should be careful not to conflate these two distinct questions.

At face value, role-holders who decide to surpass their duty and investigate the relevant first-order reasons could seriously undermine the practice in which they are participating. Hence, in contractarian terms, the parties would agree *ex ante* that under ordinary circumstances, role-holders have an obligation to abide by the rules that define their role and not waste time and energy in examining the related first-order reasons. This conclusion is strengthened given their weaker qualification to verify the relevant facts in comparison to that of the social institutes issuing the relevant orders.

On reflection, however, what could be so problematic if role-holders did take the trouble to investigate those odd cases in which they suspected that something might be wrong? Admittedly, carrying out such an investigation in all cases might undermine efficiency, but why not allow them (maybe even require them) to do so

---

[39] To be more precise, we are not interested in whether one should consider the relative weight of the relevant first-order reasons for and against the presumptively wrong action. We are interested in *practical* deliberation. As Raz emphasizes, 'exclusionary reasons are not reasons not to consider opposing reasons. Rather, exclusionary reasons are reasons not to act on opposing reasons.' See Joseph Raz, *Practical Reason and Norms* (Princeton, NJ: Princeton University Press, 1990), 39.

in a limited number of cases? We see no good reason to object to this caveat. Moreover, keeping alive the option of looking into the relevant first-order reasons might help to moderate the danger that role-holders become too loyal to their roles in a way that would erode their moral sensitivity. As a rule, then, role-holders are expected to stick to their roles and follow the instructions they receive from the relevant bodies—courts, police officers, cabinet ministers—but if, in rare cases, they wish to check the relevant first-order reasons themselves, they have a right to do so. (What ought they to do if they found that these reasons led to a different conclusion? See below.)

One might suggest that because of the interest in preserving moral sensitivity, this claim should be put in stronger terms. Instead of 'if, in rare cases, role-holders wish to check the first-order reasons, they have a right to do so', maybe the conclusion should be: 'Role-holders are obliged, once in a (long) while, to check the first-order reasons that pertain to their decisions.' But we suspect that such a rule would be rather inefficient, as role-holders would have no practical way of knowing when they are required to investigate into the first-order reasons and when they are not. Hence, it would be better to agree on a rule that grants role-holders a blanket exemption from such investigation (again, only given that the regime is in general a decent one), while acknowledging the moral benefit of them undertaking such initiatives in rare cases.[40]

Note, however, that even if there were an obligation upon role-holders to delve into the first-order merits in some cases, it would not be an obligation that granted anybody a claim against them if they did not do so in some particular case. This is because by its nature, an imperfect obligation is not owed to any particular addressee. No arrestee has a claim against the policeman who arrests him to investigate the first-order considerations that led to the arrest warrant, just as—as will become evident later—no combatant has a claim against combatants of the other side for not checking for themselves that their war is just.

What about the rare cases in which role-holders happen to know—or to believe with high certainty—that what the authorities require them to do in some specific situation is against the dictates of morality? If following the professional requirement implies knowingly violating a serious prohibition, like the killing of an innocent person, then role-holders have the right to refrain from it. True, contractarianism insists that by fulfilling their role, role-holders do not violate a right that their innocent victim has against them. They do not *wrong* their victim, nor are they responsible for the violation that the victim suffered. Yet, when they know

---

[40] As Nagel understands the moral division of labour, when individuals undertake public roles, they 'reduce their right to consider other factors, both their personal interests and the more general ones not related to the institution or their role in it' (*Mortal Questions*, 89). By contrast, in our view, such an undertaking also affects factors that *are* related to the institution of which they are part. Qua prison guard, one has a reduced obligation to consider factors relating to the inmates' guilt than one would have qua ordinary citizen.

that the serious harm that they are about to inflict is unjust, role-holders ought to refuse to collaborate with the institutions they serve (despite the fact that the responsibility for the unjust harm in question lies with these institutions). On the other hand, if the moral prohibition is not that serious (for instance, the wrongful arrest of somebody for 24 hours), then the role-related demands might take precedence.

We have focused so far on the first-order reasons that pertain to particular acts that role-holders are expected to carry out; to arrest a particular person, to execute some inmate, and so on. But there are other sorts of reasons that merit discussion; namely, reasons to believe that the regime within which one plays a role is basically decent. The moral division of labour is morally effective only when the political regime in general—and, more importantly, the particular organization of which one is part—is essentially decent.[41] If they are not, the roles assigned by the organization cannot change the distribution of natural rights and duties. In a country that is by-and-large decent, role-holders are justified in believing that the various public bodies within it—the legal system, the military, the academia, and so on—are also decent. But since it is well known that even in decent countries particular bodies might be corrupt, role-holders are expected to keep their eyes open and be alert to indications of corruption and of serious malfunctioning. If a prison guard notices that the prison manager accepts bribes from senior Mafiosi in his prison, or if he witnesses blatant lies told by the manager to his superiors, he must conclude that something is rotten in the system. Consequently, he loses the right to simply follow the orders he gets from his manager to lock inmates in their cells or to execute them.

Note that being a decent regime in the sense relevant to the present discussion is not a black or white matter. Namely, we cannot categorize a regime under the assumption that *all* of its institutional arrangements are mutually beneficial and fair or that none of them are. Rather, most regimes lie in between these extremes. Traffic policemen would retain their right to follow the rules that define their role even in societies that are generally disrespectful of human rights. The traffic laws would still determine moral rights and duties in the ways illustrated by the ACCIDENT cases. Bankers in such societies would retain the right not to play Robin Hood, and even executioners might be able to identify areas in which the legal system is doing its job properly. They would thus be justified in treating some of the verdicts of the courts as authoritative. Nonetheless, it is true that the more corrupt a regime is, the higher the chances that the corruption will have infiltrated all bodies within it; hence the stronger the (moral) expectation that role-holders

---

[41] For the claim that 'justified institutions can function as independent sources of moral obligation for those acting within them', see also Tim Dare, 'Robust Role Obligations: How Do Roles Make a Moral Difference', *Journal of Value Inquiry* 50 (2016), 715. This is a central idea in the literature on 'role ethics'.

refrain from trusting the systems to which they belong. In cases of complete corruption, that would lead to a requirement to quit one's role; namely, to refuse to collaborate with evil.

To sum up, usually role-holders have a right to disregard the first-order reasons pertaining to their actions. They may do so, however, only insofar as they are justified in believing that the system they serve is basically decent.[42] But this belief itself cannot be taken for granted; it should be reviewed and re-evaluated. If it turns out to be unreliable, the right to ignore the relevant first-order reasons loses its validity.[43]

2.7 Duels, Executions, and the Importance of Decent Regimes

We argued above that when a social rule is a necessary element in a mutually beneficial and fair arrangement, we may presume that it was freely undertaken by subjects to this arrangement. We further argued that for this reason, the rule affects the distribution of their rights and duties toward each other. In other words, fair social rules that people accept and habitually follow are morally effective, if it can be presumed that they are freely accepted. But now consider the following case:

DUEL: John makes public false accusations against Henry, which Henry denies. John challenges him to settle the matter in a duel. A refusal to fight would be interpreted as an admission of guilt on Henry's part. Henry consents to fight and shows up at the duel.[44]

---

[42] This helps to understand the fault in the claim made by people like Eichmann, pleading not guilty on the basis of 'just doing his job'. Role-holders such as officers, policemen, and prison guards have a right to treat the orders of their superiors as exclusionary reasons only if the latter are reliable in their ability to deliver better advice on issues touching on justice and on human rights. This, of course, was not the case with the Nazi regime; hence, Eichmann had no right to treat the orders he received as reasons to disregard the obvious first-order reasons against his appalling crimes.

[43] We note, furthermore, that the extent of this right depends on the nature of the role. In basically decent societies, policemen have a very broad right to follow arrest orders by their superiors without first verifying that the orders are warranted. By contrast, officials providing welfare services must not rely only on the procedures and protocols of their bureau, but must also show sensitivity to the unique features of the cases they deal with, to prevent tragedies of the kind described in the 2016 film *I, Daniel Blake*.

[44] This is a revised version of an objection raised by McMahan, *Killing in War*, 56. See the discussion in Benbaji 'The Moral Power of Soldiers to Undertake the Duty of Obedience'. Compare to Tadros, who rejects the view that combatants have a right to harm each other because they consented to being liable to be harmed in the course of a war. In his view, 'consent is neither necessary nor sufficient to create a conflict of permissibility' (*The Ends of Harm*, 114). However, at least according to our understanding, consent *in and of itself* does not do the normative work we assign to it; other conditions must be assumed as well.

2.7 DUELS, EXECUTIONS, AND DECENT REGIMES 63

Does Henry's consent release John from the duty not to kill Henry? Does Henry lose his right to life vis-à-vis John, just like prisoners in the death row do vis-à-vis their executioners, arrestees vis-à-vis the policemen who arrest them, and so on? One shies away from giving a positive answer. Why? If *ex ante* consent of the sort expressed by habitual obedience to social rules explains the moral effectiveness of social rules, actual consent of the kind given by John and Henry should do so all the more.

The immediate answer seems to be that *Mutual Benefit* is not satisfied in this case, so the consent should not be understood as free in the relevant sense. There are many better ways of resolving disputes between people. Duels likely amount to one of the worst conflict resolution methods. Hence, it seems that Henry's showing up at the duel does not provide John with a moral permission to kill him: his consent to this arrangement is coerced by an oppressive social arrangement. If John does kill Henry, John would be wronging him.

However, imagine a society in which duels are the only effective mechanism for maintaining social order. The regular law-enforcing bodies are impotent or non-existent. In this scenario, it probably would be rational for members of society to agree *ex ante* on the rules governing the practice of duelling. The risk they would undertake by agreeing to such rules would be smaller than that involved in living in a society that had no accepted way of resolving disputes. True, this system lets the physically stronger party have its way; yet it is *ex ante* better for all parties, including the weak ones, than a system in which there is no agreed-upon way to resolve conflicts.

Admittedly, such a society is quite imaginary. Historically speaking, duels were always an additional way of settling disputes—usually disputes concerning honour—and not a substitute for a functioning legal system. Moreover, societies usually depended on the existence of such a legal system to regulate and enforce the practice of duels. Yet, as a thought experiment, this imaginary society is nevertheless helpful in challenging the normative picture we construed in previous sections. It seems that even if the practice of duels was (*ex ante*) mutually beneficial to all members of society, and even when particular members arrived at the duel scene voluntarily, killing in a duel would be wronging the victim—which seems to run against the contractarian view proposed above.

There is however a crucial difference between the regime described by DUEL and those described in BANKER, POLICEMAN, and EXECUTIONER. Consider the regime where the role of the executioner is institutionalized. If the arrangement is fair, individuals lose their right against executioners not to be put to death by them. But they do not lose their right not to be unjustly killed by the relevant social institutions, mainly courts. Within such a regime, responsibility is transferred (so to say) from the person carrying out the killing (namely, the executioner) to the social institutions that ordered him to do so (i.e. the court). The victim has no claim against the executioner not to kill him, precisely because, within the conventionally sustained social structure, he has a claim against the state (or the courts, or the judges) not to be executed for morally invalid reasons.

In these respects, DUEL is significantly different from EXECUTIONER. In the society we are imagining, John would be acting as a private individual advancing his own interests. There is no recognized authority that he could fall back on to justify his attack on Henry in the duel. There is no other authority to which the responsibility for not violating Henry's right would be transferred. *A fortiori*, there is no entity that could be relied upon to decide in questions regarding justice and rights. By contrast, in a fair and well-functioning regime, the responsibility for the execution of assumed criminals is transferred from the executioner carrying out the killing to the social institutions that ordered him to do so. Although the executioner is the immediate cause of the criminal's death, the responsibility for this outcome lies fully on the shoulders of the state. The executioner is morally responsible only for the technical and humane aspects of the execution, not for the decision to execute.[45]

Since people governed by a fair and well-functioning regime agree to the role of the executioner, they lose their right against executioners not to be unjustly executed by them, just as they lose their right against policemen not to be unjustly arrested. But these losses do not affect the responsibility to protect these rights, which is incumbent on other role-holders within the system. Duels express a different reality. If John unjustly kills Henry, it is John who would be wronging him, and it is John against whom Henry would have a claim-right not to be attacked.

We suggest generalizing this result under what we shall call the 'law of conservation of rights violations' (hereafter, *Conservation*). If a decent regime allows one person to harm another unjustly, there is necessarily some individual or institution who is responsible for this violation; some role-holder (judge) or public body (court) that could rightly be said to have violated the victim's right. We insist, though, that the violator need not be the role holder who actually harmed the victim. In accepting a decent regime, people trade with their rights, giving them up in exchange for various benefits. When the regime is decent, they give them up only vis-à-vis specific people; they freely accept a regime that transfers the responsibility from the immediate cause of the rights violation to another role-holder in the social structure. Thus, as we shall see later, while combatants lose their right not to be killed by combatants of the other side, they don't lose it against the politicians who sent those enemy combatants to the battlefield.

The point could also be put in terms of reasonable complaints. If I trade my right not to be wronged by a particular individual or individuals, in a certain way, under well-defined circumstances, I still retain my right to complain against

---

[45] See Applbaum, *Ethics for Adversaries*, Chapter 2 ('Professional Detachment: The Executioner of Paris').

others who are responsible for such wronging. Prisoners who are unjustly imprisoned have no cause for complaint against their guards for locking them in their cells, but they do have a right to complain against the people who ordered their guards to unjustly imprison them. And while a combatant has no cause for complaint against the combatants of the other side who fire at her, she does have a cause for complaint against the political leaders of the enemy who decided to launch (what she sees as) an unjust war.

Why are violations of rights to life and bodily integrity subject to *Conservation*? The short answer is that since we have human rights 'simply in virtue of being human',[46] we cannot lose them so far as we remain human. All we can do is to transfer the responsibility to respect them, which means giving up our moral immunity vis-à-vis specific agents in specific circumstances in exchange for perceived benefits. It is not within our power to give up this immunity altogether, because that would express a denial of our moral status as bearers of rights. A fuller answer requires the development of a comprehensive theory of human rights, which extends beyond the scope of the present project.

Note that *Conservation* leaves everything open regarding the question of what measures, if any, should be taken against the individuals or the social and political institutions that bear responsibility for the violation of rights. In particular, it does not follow from what was just said that they must face criminal charges for such violation. In decent regimes, judges should not face the charge of homicide in case they unjustly sentence a person to death. How to respond to such violations of rights depends on many factors, and we need not settle the issue here.

As stressed above, the division of labour in society is effective only if the society's institutions are basically decent. The present section illuminated a central implication of this condition that is often overlooked; namely, that these institutions are responsible for violations of rights carried out by individuals who are subject to them. A regime in which the relevant institutions care about and take responsibility for violations of rights is morally preferable to the DUEL regime. Although the DUEL regime meets the conditions of *Mutual Benefit* and *Fairness*, it is deficient in terms of respectfulness for human rights. Such a regime lacks institutions that care about and that take responsibility for the unjust death of people in duels.

Note, finally, that *Conservation* does not apply to the case of boxing discussed above. Boxers are allowed to punch each other in the ring because each has relinquished his right not to be hit (vis-à-vis the other boxer, in accordance with the rules of the game, and so on). But unlike POLICEMAN or EXECUTIONER, in this case there is no remainder of the relinquished right; it is not conversed in any other form, or towards any other agent. Probably, the difference between boxing

---

[46] James Griffin, *On Human Rights* (Oxford: Oxford University Press, 2008), 2.

and these other cases follows from the nature of the rights under discussion. Some moral duties and their correlative rights are almost inalienable. The duty not to kill innocent people is binding even if the potential victims consent to being shot. That is why one cannot waive one's right not to be killed in a duel or in a game of Russian roulette. Henry's voluntarily showing up at the duel does not absolve John from the wrongness of killing him. This is one of the limits of our normative power to exempt others from their duties toward us.[47] In contrast, the natural right not to be touched or punched or even wounded is alienable and *can* be waived by boxers in exchange for various benefits.[48]

2.8 Between Contractarianism and Rule-Consequentialism

According to contractarianism, individuals living in a decent society ought to or are allowed to follow their society's rules, as long as these rules are mutually beneficial and fair. The moral standing of specific acts is sensitive to the extent to which they conform to or divert from such rules. This understanding of contractarianism seems close to rule-consequentialism (RC), which also places rules at the centre of the moral landscape and assesses their moral validity by exploring the benefit that is expected to ensue from keeping them. Brad Hooker, the influential defender of RC, articulates the role of rules in this theory as follows:

> The theory [RC] can be broken down into (i) a principle about which rules are optimal, and (ii) a principle about which acts are permissible. The theory selects *rules* by whether their internalization could reasonably be expected to maximize the good. The theory does not, however, evaluate *acts* this way. Rather, it evaluates acts by reference to the rules thus selected.[49]

More importantly, like RC, our version of contractarianism relates moral rules to the expected benefit of following them rather than—*a la* Scanlon—to the mutual recognition that is expressed in their acceptance.[50] As Hooker puts it, like RC, 'Hobbesian contractarianism draws on an extremely appealing idea—that morality is a system of mutually beneficial co-operation'.[51] These points of similarity raise the question of what distinguishes contractarianism, as we understand it,

---

[47] Tadros, *The Ends of Harm*, 214.
[48] For the sake of argument, we assume that the risk of boxing is relatively minor ('to be punched in the nose'). But the actual risk is probably much higher, to the extent that it undermines the claim that the practice is *ex ante* mutually beneficial to all involved parties and, as a result, also undermines the moral permission to participate in it while disregarding the natural rights of one's adversary.
[49] Brad Hooker, *Ideal Code, Real World: Rule-Consequentialist Theory of Morality* (Oxford: Oxford University Press, 2000), 102.
[50] Scanlon, *What We Owe to Each Other*, 162.
[51] Hooker, *Ideal Code, Real World: Rule-Consequentialist Theory of Morality*, 7.

from RC. The following four points are meant to provide an initial answer to this question. A full answer requires a much deeper investigation into these two theoretical outlooks, a task that lies beyond the scope of the present book.

First, while RC offers a theory about the nature of wrongness and rightness of actions, which relates these features of actions to the rules that allow or prohibit them, (our version of) contractarianism does not offer a theory about the nature of morality—and in particular, it offers no theory about the basis of fundamental moral rights. Rather, it *assumes* the existence of pre-contractual rights and argues that they might be distributed (or redistributed) by social arrangements. In other words, what contractarianism has to say about the nature of morality is precisely that social arrangements might affect the distribution of fundamental moral rights.

Second, and relatedly, most versions of RC (and of consequentialism more generally) do not treat rights as fundamental to morality. At best, they derive them from consequentialist considerations. At worst, they deny that they exist. By contrast, in our view, rights—which we refer to as 'natural' or as 'pre-contractual'—are fundamental and serve as the building blocks of ethical theory. On this point, our Lockean version of contractarianism is inconsistent with the contractarian morality developed by Hobbes and Gauthier. Hobbesian morality states that the social contract does not entail a mere distribution of moral rights but rather *creates* them *ex nihilo*.[52]

Third, in contrast to RC, from a contractarian perspective the effectiveness of the rules does not depend on any notion of maximization. The rules need not maximize benefit; rather, they must be *mutually* beneficial. They are designed such that they are expected to benefit *all* relevant parties. As such, habitual obedience to them brings about a state of affairs that is expected to be Pareto superior to a state of affairs in which parties are subject to and try to follow pre-contractual morality.[53]

Fourth, the crucial difference between contractarianism and RC may be that under the latter, being (mutually) beneficial is not what makes the social arrangements morally effective. Rather, the fact that rules are mutually beneficial validates the presumption that they are freely (as well as tacitly) accepted, and the fact that they are tacitly accepted entails a waiver of rights. In other words, rules that are mutually beneficial and are habitually followed are presumably rules freely undertaken. Thus, while under RC, violations of rights that result from the moral

---

[52] This seems to be Applbaum's view as well (suggesting that 'morality is, at bottom, simply about the summing up of benefits and burdens across persons', *Ethics for Adversaries*, 13). However, in other respects, we are close to his view; i.e., in his 'contractualist sensibility' (*Ethics for Adversaries*, 13) and his interrelated acceptance of *Moral Division*.

[53] The relation between Pareto efficiency and utilitarianism is more complex, but treatment of this issue lies beyond the scope of this discussion. See, for example, Powers Madison, 'Efficiency, Autonomy, and Communal Values in Health Care', *Yale Law and Policy Review* 10/2 (1992), 316–61.

division of labour are justified on the basis of their (direct or indirect) contribution to the overall good, under contractarianism, when role-holders harm people while legitimately doing their job, they ultimately do not violate the rights of their victims because the latter tacitly accepted the regime that allows role-holders to bring about such harms.

2.9 Conclusion

In the previous chapter, we became aware of the spell that *Individualism* has woven on the philosophical discussion about the ethics of war in the last two decades or so, and we elaborated on the problematic implications of this spell. The purpose of the current chapter is to outline an alternative to *Individualism* and to show that one can take moral rights very seriously while acknowledging the role of organized societies in determining the actual distribution of rights and duties of citizens. In some cases, the rules accepted by such societies give content to what was indeterminate at the pre-contractual level. In others, they re-distribute moral rights and duties among members of society. In both these ways, rights behave in a less rigid manner than that entailed by *Individualism*.

To understand how social rules can determine rights, it is particularly helpful to look at the way social roles provide their holders with a permission to diverge from what would be required from them pre-contractually. In decent societies, policemen, deans, bankers and other holders of public roles typically have an exclusionary reason to fulfil their professional duty, without deliberating on the merits of the case; namely, without being guided by first-order reasons that pertain to the cases with which they deal. Moreover (and in this regard, contractarianism goes far beyond the service view), the fact that these authoritative rules are accepted in society implies a distribution of rights and duties in this society. The idea that role-holders have a right to divert from pre-contractual morality is essential to understanding the moral status of combatants.

The approach of this book is rooted in the social contract tradition. As this tradition is rich and varied, we would like to underscore two points concerning the relation between our understanding of it, and that of others. First, contrary to the Hobbesian approach, we assume that individuals have pre-contractual rights—especially to life and to bodily integrity—which are independent of any political framework. In this sense, the contractarian framework by which we argue for the traditional war theory is Lockean. Second, unlike thinkers like Rawls and Scanlon, we do not deduce the binding force of social rules from the fact that rational and self-interested people behind a veil of ignorance would accept them, but from the fact that they satisfy the conditions of *Mutual Benefit, Fairness* and *Actuality*. In the contractarian framework developed here, the actual acceptance of the rules is a crucial aspect of the moral validity of norms. Thus, *Actuality* is especially worth emphasizing: social rules are morally binding only if they are actually adhered to (or *accepted*, in Hart's terminology); which, when the other conditions are met, indicates free acceptance of social rules and a waiver of rights.

With regard to this free acceptance, contractarianism makes another assumption: By freely accepting a legal system, citizens authorize the state (that institutionalizes this system and enforces it) to act on their behalf in the international realm. Hence, if it is fair and mutually beneficial, a treaty-based international law is as morally effective as is the domestic law by which these individuals are governed. While the war agreement is best understood as a treaty-based international law, it is morally effective only if decent individuals would have entered it themselves. It is morally effective only if, in signing it, states acted on the behalf of their (self-interested and yet decent) citizens, such that under this war agreement, these citizens enjoy better security, and their rights are better protected.

It might be helpful to summarize the normative framework that we employ in this book thus:

a. Individuals have rights independently of any political framework (which we refer to as 'pre-contractual' or as 'natural' rights). Any political society to which these individuals belong is obliged to respect these rights—in particular the rights to life and to bodily integrity.

b. Individuals have a weighty prudential reason, and probably (as Kant argued) also a moral reason to be citizens of a decent state, i.e., to move from the state of nature to the civic condition.

The other two assumptions are inspired by Hart's theory of legality:

c. A group of people forms a political society only if the group is governed by a legal system that is composed of social rules.

d. Rules are social by virtue of being 'accepted', namely, being adhered to in practice within the society in question.

Contractarianism loads the Hartian concept of acceptance with normative significance by relying on two further assumptions. The first presents the conditions under which the acceptance of rules can be seen as free.

e. Presumably, a system of social rules is accepted freely by those who subject themselves to it if and only if it is sufficiently good.
  1. A system of legal rules is sufficiently good only if it is mutually beneficial; viz., its acceptance is likely to bring about a Pareto-superior outcome compared to an outcome in which most try to fulfill their pre-contractual duties and take advantage of their pre-contractual rights.

2. A system of legal rules is sufficiently good only if it is fair; namely, it neither creates nor solidifies unfair or disrespectful social relationships.

Now,

f. By freely accepting a legal system, individuals implicitly consent to be governed by it. Consequently, citizens waive some of their pre-contractual rights. For example, by undertaking the duty not to use force against each other, citizens waive their pre-contractual right to defend themselves by force against violations of their rights, insofar as the state can provide such a defence for them.

In order to explain the effect of the war agreement on the distribution of individual rights, contractarianism makes a further assumption:

g. By freely accepting a legal system, citizens authorize their state to act on their behalf in the international realm. Hence, if fair and mutually beneficial, a treaty-based international law is as morally effective as the domestic law by which these individuals are governed.

In concluding this chapter, we should be clear about its modest aspirations. Clearly, one cannot develop a complete ethical theory in one chapter; and definitely, one cannot provide a conclusive argument for its advantages over other theories. We do not presume to have done so. What we hope to have achieved is to motivate a distinct moral perspective that explains widespread intuitions in a range of cases. The approach sketched here, we hope, coheres well with a plausible version of contractarianism. A full defence of this approach is not a task that we can undertake here, but we hope that our explication and defence of non-individualist, contractarian morality will prove helpful when we turn to the ethics of war. This will be our concern in the following chapters.

3
A Contractarian Account of the Crime of Aggression

3.1 Introduction

The purpose of this chapter is to start explaining the moral standing of *jus ad bellum* as it is formulated in the UN Charter, on the basis of the contractarian framework established in the previous chapter. The basic idea is that the acceptance of the Charter affects the distribution of the rights of states (and of non-state actors [NSAs]) to use force. Any armed violation of a state's territorial integrity by another state is an instance of prohibited aggression. The contract confers a right against aggression even on states whose borders are unjustly drawn, and even on dangerous states whose political society is irrecoverably divided. The acceptance of the Charter is morally effective because it embodies a good (perhaps even an optimal) contract to which states have subjected themselves. By accepting it, the parties waive their right to go to some pre-contractually just wars (like subsistence wars, preventive wars, or wars whose aim is a just regime change), but gain a right to go to some pre-contractually unjust wars (like wars whose aim is the maintenance of territorial integrity).

Does the war agreement obligate individuals although it is signed by (their) states? As we noted earlier, contractarianism assumed that it does. By freely accepting a legal system, citizens authorize the state that enforces this legal system to act on their behalf in the international realm. Therefore, the war agreement is morally effective only if, in signing it, states act on behalf of their citizens. In particular, the war agreement determines the distribution of moral rights and duties, only if, thanks to its existence, individuals are safer and their rights better protected. The implicit authorization that citizens grant their states to reach a war agreement is presumably free if the rules which states draft satisfy the conditions of *Mutual Benefit*, *Fairness*, and *Actuality*.

Most of our discussion revolves around the *ad bellum* condition of just cause, demonstrating why the protection of sovereignty and territorial integrity is rightly considered the only *casus belli*.[1] However, contractarianism sheds light on the other conditions for *jus ad bellum* too, mainly necessity ('last resort') and

---

[1] Below, we argue that the protection of these two values—territorial integrity and sovereignty—should be read as covering more or less the same field.

proportionality. We start our discussion of these in the present chapter and elaborate further in the following one.[2]

3.2 Why the *Ad Bellum* Regime is Mutually Beneficial

We have presumed that parties' acceptance of a legal system is morally effective; viz., it entails a waiver of rights, as long as the arrangement that the system enforces is mutually beneficial and fair. In what follows, we show that if certain empirical assumptions are true, the *ad bellum* regime satisfies these conditions and, therefore, its acceptance by the relevant parties is morally effective. We assume that the Charter is mutually beneficial in the relevant sense if and only if (a) partial, decent states prefer an outcome in which the contract is accepted by the overwhelming majority of states to an outcome in which such states are governed by principles of pre-contractual morality; and (b) it includes rules that deter parties from violating the contract and from being or becoming indecent.

States are partial in three distinct senses. First, they prefer the promotion of their own interests to the promotion of the interests of other states. Second, they care about protecting their own rights more than they care about protecting the rights of others. Finally, partial states are biased towards themselves. They tend to judge the normative and factual issues at stake in a way that is consistent with their own interests.

The partiality of states means that the threat of armed conflicts between decent states comprises a permanent element of international life. In the absence of a global government that could enforce the rights and entitlements of states, states are constantly on guard to provide security to their citizens. Given the prudent suspicion that they have towards other players, the danger of explosion is real.

Turn now to decency. The model deals with decent states (which should resonate with 'decent peoples' in Rawls's political philosophy[3]). Despite their partiality, such states acknowledge that individuals are subjects of rights and that each other member state within international society is entitled to sovereignty. States are decent in another sense: they tend to respect the contractual duties they undertake. The fact that they placed themselves under a contractual duty amounts to a weighty reason for them to respect it.

A world composed mostly of decent but partial states with no common authority to adjudicate conflicts between them constitutes what we shall call 'a minimally just anarchy'.[4] It is an 'anarchy' because of the absence of such common

---

[2] The following sections draw on Yitzhak Benbaji, 'A Contractarian Account of the Crime of Aggression', in Fabre and Lazar (eds.), *The Morality of Defensive War*, 159–84.
[3] John Rawls, *The Law of Peoples* (Cambridge, MA: Harvard University Press, 1999), 37.
[4] The phrase is Hedley Bull's. See his *The Anarchical Society: A Study of Order in World Politics* (Oxford: Oxford University Press, 1977).

authority. It is 'minimally just' because thanks to the minimal decency of the parties, most of the perceived injustices in international relations are only moderately unjust in the sense that an impartial observer will find it hard to determine whether it is morally justified to remove the injustice by using lethal force. This anarchical state of affairs is also 'symmetrical' if in most conflicts even the strong party has a prudential reason to solve it by bargaining rather than fighting.

Given these features of the international reality, we set out to show that states have a solid reason based on both self-interest and morality to undertake an arrangement that denies the right to wage most pre-contractually just wars. While pre-contractual morality might license justice-implementing wars—be they subsistence wars, preventive wars, or wars of humanitarian intervention—such wars are ruled out by the agreed-upon regime because states waive their right to wage them. Hence, the over-restrictiveness of Article 2(4) is not troublesome. Neither is the over-permissiveness of Article 51. Although states might have no pre-contractual right to wage wars of national defence against violators of their territorial integrity, the parties reach a contract that grants them such a right—hence Article 51.

Because of states' partiality, they are sometimes misguided about the facts or about their moral significance. Conjoined to their partiality, their ignorance might bring about armed conflicts between states; decent states could find themselves in armed conflicts with other decent states. As we shall immediately argue, to minimize armed conflicts that under symmetrical circumstances are mostly bad for all parties, states should undertake a sweeping prohibition against first use of force. If such a rule were followed, that would create a better world in terms of both the promotion of the parties' interests and the protection of human rights.

Importantly, we read the war agreement as having another crucial function—incentivizing states to undertake the Article 2(4) prohibition and deterring them from violating it. Obviously, not all states are decent. Some individuals and some states care very little about morality. Others do care about morality, but are too partial. The war agreement must include rules of two types: rules of the first type enable states to deter other states from becoming indecent; whereas rules of the second type enable them to handle the threats posed by rogue states that are thoroughly indecent. As Hooker puts it, when we think about social rules, we should think of 'what is needed to deter or rehabilitate someone who has a moral conscience and accepts the right rules but sometimes does not care enough about morality to ensure good behavior', alongside 'what is needed to deal with unmitigated amoralists (people who have no moral conscience at all)'.[5]

---

[5] Brad Hooker, *Ideal Code, Real World: Rule-Consequentialist Theory of Morality* (Oxford: Oxford University Press, 2000), 82.

In this chapter we deal with rules of the first type; viz., rules for using force against other decent parties. We explore rules governing use of force against indecent states and indecent NSAs in Chapter 7.

We turn now to describe in more detail the circumstances of symmetrical anarchy against which the Charter is expected to benefit decent states that find themselves in conflict.[6] In a situation of symmetrical anarchy, for almost every future conflict, C, between partial and decent states, there is at least one peaceful resolution that is Pareto superior to a state of war. To see how this condition might be satisfied, consider conflicts of interests over divisible and commensurable goods. Assume that these conflicts can be represented by cardinals, just like domestic disagreement over a set of issues whose market value is, say, 10,000.[7] The parties in such conflicts can either bargain or fight, like individuals in a domestic society who can bargain or go to court to resolve their conflicts.

Now, under *a*symmetrical circumstances, the probability that the strong party ('Strong') would defeat the weak party ('Weak') in future armed conflicts is so high, while the cost it will have to bear for its using force is so low, that Strong's use of force is its dominant strategy. Whatever Weak does, fighting would be in the best interest of Strong. In contrast, under symmetrical circumstances, most conflicts are such that a peaceful compromise is preferable *ex ante* to fighting for both parties. Here is a simple illustration. Suppose that the probability that Weak will win the war is 0.3, while the probability of Strong's winning is 0.7. Suppose further that the cost of the war for Weak is 2,000. This means that Weak's expected benefit from the war is 1,000.[8] Therefore, any compromise or peaceful resolution of the conflict under which Weak accepts more than 1,000 is better for Weak than going to war. Suppose now that the cost of the war for Strong is 3,000.[9] Hence, the expected benefit of its war against Weak is 4,000. Therefore, any compromise under which Weak gets more than 1,000 and Strong gets more than 4,000 is *ex ante* preferable to both parties than war. Following James Fearon, we shall say that any agreement in which Weak gets more than 1,000 and Strong more than 4,000 belongs to the 'bargaining range' (defined as the spread between the resistance points of the two parties) of the conflict.

Put differently, each party has a strategic set of options comprised of two elements: {fight, bargain}. The outcome <fight, fight>—where the conflict is resolved by fighting—is Pareto inferior to <bargain, bargain>, where the conflict is resolved by bargaining.

We thus define symmetrical circumstances through the following seven conditions:

---

[6] The analysis is borrowed from James Fearon, 'Rationalist Explanations for War', *International Organization* 49 (1995), 379–414.
[7] This assumes that the issues have the same value to both parties. This is a strong simplifying assumption, which is unlikely to be true in reality. We shall relax it later.
[8] 0.3.10,000=3,000; 3,000 minus the costs of the war is 1,000.
[9] 0.7.10,000=7,000; 7,000 minus the costs of the war is 4,000.

3.2 WHY THE *AD BELLUM* REGIME IS MUTUALLY BENEFICIAL

(a) The basic condition: most conflicts that decent states face under these circumstances have a positive bargaining range.

(b) States are aware of the basic condition; namely, they know that in most future conflicts, bargaining will yield an outcome that both parties would find preferable to fighting.

Both conditions seem realistic. First, wars are so costly in so many respects that peaceful bargaining is almost always *ex ante* preferable to fighting, and thus is clearly mutually beneficial to all relevant parties. (The exception would be wars aimed at mass murder or severe oppression, in which the attacked party would prefer the <fight, fight> option.[10]) Second, given the history of wars in general, and often also those in which the particular parties have been involved, states usually realize that for most conflicts, <fight, fight> is Pareto inferior to <bargain, bargain> in terms of expected benefits assessed before the war starts.

The third condition comes from Fearon's game-theoretic explanation for why states go to war in order to resolve symmetrical conflicts. Even if a conflict between two states has a positive bargaining range, rationally-led states might resolve it by going to (Pareto-inferior) wars. This is because they have an incentive to present themselves as more powerful than they actually are and to present their use of force as less costly to them than it actually is. The obvious reason is that a successful bluff might secure a better deal for the bluffing party. If Weak is able to convince Strong that its chance of winning is 0.4, Strong would offer a compromise under which Weak gets more than 2,000. Thus, in symmetrical conflicts, both parties have a reason to suspect that the other side is bluffing. This suspicion might lead them to erroneously believe that the conflict they face has no bargaining range, and that *ex ante*, fighting is preferable to any feasible peaceful resolution. Thus, lack of information might push states to war even if, had they had the relevant information, they would have bargained rather than fought. Thus, the third condition:

(c) States under symmetrical circumstances are aware of the danger that lack of information might cause them to go to war against each other.

The other features of symmetrical circumstances involve the nature of the treaty-based regime that rationally-led states would adopt in order to avoid Pareto-inferior wars. The fourth condition follows immediately from the first three:

(d) Because, usually, fighting is Pareto inferior to bargaining, it is in states' self-interest to enter a contract that condemns first use of force, if and only if other parties join the treaty and (for the most part) observe it.

---

[10] In those cases, one of the parties is indecent. We discuss such cases in Chapter 7.

In particular, it is in their self-interest to enter a contract that prohibits pre-contractually just wars if others join it and if the contract is actually observed. Such a contract advances their interests in light of the statistical fact expressed in the basic condition (a).[11]

The final features of symmetrical anarchy are supposed to explain why if bargaining is almost always Pareto-superior to fighting, decent states would not commit themselves to pacifism. Why wouldn't it be mutually beneficial for them to undertake an agreement that rules out wars altogether? The answer appeals to problems of commitment and of collective action, which render a regime that outlaws all uses of force unattainable in practice. Because each state is interested, first and foremost, in protecting its own interests, it would like everyone *else* to be committed to pacifism, while it alone retains its right to go to pre-contractually just wars. This situation would enable the state to cheaply and easily enforce what it takes to be its rights. In the absence of a universally recognized authority to ensure that all parties respect their commitment, a pacifist contract would be unworkable.

Therefore, we suggest that symmetrical anarchy also satisfies the following conditions:

(e) Condition (d) is likely to be satisfied only if the contractual duties that the agreement contains—especially the contractual duty not to wage pre-contractually just wars—are enforceable.

Alas,

(f) The society of states lacks a central government; hence, contractual rights/duties conferred by an agreement can be enforced by self-help only.

And,

(g) The self-help based regime to which states subject themselves in order to avoid inefficient wars would treat any violation of territorial integrity as a just cause for war (unless the contract explicitly allows the violation in question). The territory of a state is also invaded when the enemy aircraft enter its airspace in order to carry out hostile activities, when enemy warships invade its territorial waters, and, of course, when its enemy fires missiles into state territory.

The self-help clause is implemented through a right to go to defensive wars against states that use force in violation of the contract—a valid right regardless of whether or not the threats in question are pre-contractually justified.

---

[11] States have a further reason to enter this contract. Errors in assessing the costliness of force or the probability of victory are common; rule-governed wars might very quickly become total and hence much costlier than initially anticipated.

Defining national defence and the crime of aggression through territorial integrity has a clear rationale in symmetrical anarchy. Under such circumstances, most wars that states try to avoid by entering the agreement involve imminent threats to the territorial integrity of states, and vice versa: most violations of territorial integrity are a link in a chain of events that ends up leading to inefficient wars. Moreover, threats to territorial integrity are the most visible link in the chains of events that lead to Pareto-inferior wars. By defining aggression in terms of (threat to) territorial integrity, the Charter provides the simplest and most efficient way to enforce the *ad bellum* contract and thus to minimize the occurrence of Pareto-inferior wars.

The reasons we mentioned for regarding (imminent) threats to territorial integrity as the only *casus belli* also explain why threats to sovereignty should be interpreted in a very restrictive way, usually as threats to territorial integrity. A wider reading of the notion would grant much too broad a permission to wage war in assumed protection of the state's economic interests, its international relations, or its domestic policies. Thus, the only threat to sovereignty that justifies use of force by an individual state is a threat to its territorial integrity.

However, the restriction of the right to wage war to circumstances involving a threat to territorial integrity does not explain a further restriction on going to war; namely, the requirement of imminence. If your enemy is planning an invasion, why wait until the threat is imminent, when defence might be harder? Why not launch a *preventive* war weeks or maybe months before that point is reached?

To appreciate this problem, consider the case of what Posner and Sykes called an 'optimal' preventive war. A preventive war, W, is optimal if the total expected harm (to innocents) caused by W is smaller than the harm that a permissible defensive war W* would cause, multiplied by the probability that W* will erupt. Contractarianism must explain how a code that permits W* could prohibit W, if W prevents W*, and if W is less costly (in terms of both utility and violation of rights) than W*. Shouldn't states under symmetrical anarchy agree on a rule that permits wars that enforce the contractual duty not to use force (W*s), as well as optimal wars (Ws) whose goal is the prevention of W*s with less human and economic distress to both sides?[12]

Indeed, the problem applies to the pre-contractual right to self-defence as well as to contractually just wars. Usually, a private person has the right to use force against an aggressor only if the threat that the aggressor poses is imminent. An explanation is needed for the demand to wait for the threat to mature. Similarly, why is the right to self-defence conferred by the war agreement restricted to imminent threats?

---

[12] See Eric Posner and Alan Sykes, 'Optimal War and *Jus Ad Bellum*', *Georgetown Law Journal* 93 (2005), 993–1015.

We believe that contractarianism offers a simple explanation for the restricted scope of the contractual right against aggression. The aim of the contract embodied in the Charter is to reduce the impact of the information problems that cause inefficient wars. Therefore, the enforcement rule contained in the war agreement must pass the following epistemic test: its application should not cause further significant uncertainties. Threats that the Charter defines as just causes for war—that is, imminent threats to the territorial integrity of states—pass this test. In contrast, the information needed in order to determine whether a war is optimal is hard to obtain. This type of determination relies on probability assessments about future events that are unstable and vulnerable to various biases. Such uncertainty in itself encourages unethical behaviour, as empirical research shows.[13] Given these facts, partial yet decent parties should agree upon rules that are as clear and as enforceable as possible. The rule that war may be waged only in the face of imminent threats passes this test.

The contractual prohibition on preventive wars has an important yet troubling implication. Consider the following scenario: State A is developing a threat to State B that will not come to fruition for some time but is certain to be devastating. B is denied the right to fight an optimal war to avert that threat. However, B does so anyway. As a result, A has the contractual right to use force against B. This means that A could precipitate a (contractual) just cause to fight by developing a pre-contractually unjust threat to an enemy.

Contractarianism should bite this bullet. In light of the possibility of a justified violation of the duty imposed by the contract, such a scenario is not that implausible; the regime is contractually justified by virtue of the benefits that states gain from it in the long run. As argued several times above, the fact that *ex post* I am about to lose by keeping some agreement does not exempt me for my duty to respect it. Notably though, this duty is not absolute. In rare cases, states might be justified in violating their contractual duty not to use force, just as executioners or policemen might be.

---

[13] Thus, Doyle argues that 'subjective and abstract standards of prevention are ... much too likely to be self-serving, promoting narrow partisan advantages' (Michael W. Doyle, *Striking First Preemption and Prevention in International Conflict* (Princeton, NJ: Princeton University Press, 2008), 29). Accordingly, Doyle sets out 'four standards for anticipatory self-defense' that would help avoid such problems: 'Lethality identifies the likely loss of life if the threat is not eliminated; Likelihood assesses the probability that the threat will occur; Legitimacy covers the traditional just war criteria of proportionality, necessity, and deliberativeness of proposed responses; and Legality' (46). Cf. David Luban, 'Preventive War', *Philosophy and Public Affairs* 32 (2004), 207–48. Luban argues that 'no evidence can show that a ban on preventive war would save lives'; this, he believes, does not undermine the importance of the prohibition. For, 'no evidence can show that *any* doctrine of just war saves lives, simply because states so frequently disregard moral and legal norms'. For Luban, 'the right test for a moral norm should not be whether the norm will be efficacious, but rather whether it would be efficacious if states generally complied with it' (226). See, on the other hand, Max H. Bazerman and Ann E. Tenbrunsel, *Blind Spots: Why We Fail to Do What's Right and What to Do About It* (Princeton, NJ: Princeton University Press, 2011), 164: 'The more uncertainty there is in the environment, the more likely unethical behavior is to occur.' We elaborate further on this point in Section 6.2.

Our analysis of the *ad bellum* regime raises two difficulties. First, if states in conflict C avoid fighting because they realize that resolving the conflict by <fight, fight> is Pareto inferior to resolving it by <bargain, bargain>, then the law that prohibits the use of force is redundant. Plain self-interest would lead states to refrain from war. If, however, states in C cannot cooperate because of information problems, then they *will* fight, whether the legal system condemns wars or not; the *ex ante* agreement will be ignored *ex post*. Either way, the legal regime that restricts the use of force would not have any real impact on the behaviour of states.

Indeed, this is a difficulty for contractarian accounts in general. The *ex ante* mutually beneficial nature of some arrangement can explain why the parties sign on to it, but it seems to fall short of explaining when they would comply with it *ex post* if they believe that such compliance runs against their interests. Note that this is not a question about moral justification but about moral motivation (or motivation to follow morality). Against some background conditions, a free acceptance of some contract surely imposes a legal and moral duty upon the accepter to follow the terms therein. The question is a psychological one; how can parties ignore their self-interest and abide by their contractual duties?

A full answer to this fundamental question about human nature is beyond the scope of the present study. Suffice it to say that in our view, Hume was right in assuming that human beings are capable of transcending their narrow self-interest and taking into account the interests of others. Joshua Greene recently pointed to various ways in which our psychology evolved to make us sensitive to the pain of others and loyal to the commitments we undertake.[14] Hence, our psychology does not make it impossible to respect our commitments in case we realize that doing so would go against our self-interest. In addition, the *ad bellum* contract includes not only a prohibition against the first use of force but also a way of *enforcing* this prohibition; namely, a right granted to states to launch defensive wars in order to enforce their contractual right not to be attacked. This right has a deterring force, which is supposed to make states think twice before they succumb to their narrow self-interest and violate their contractual commitments.

The tendency to comply with the war agreement is strengthened by another empirical fact. Politicians care a lot about public sentiment when making critical decisions. In decent states, the public tends to weigh international legal agreements positively. Hence, politicians will often find it hard to violate such agreements, given their fear of losing the support of their public.

Consider another important difficulty faced by the argument we have just developed: suppose a state considers going to a pre-contractually just war in one

---

[14] Joshua Greene, *Moral Tribes: Emotion, Reason, and the Gap Between Us and Them* (London: Atlantic Books, 2015), Chapter 2.

of the rare cases where fighting is not Pareto-inferior to bargaining. Is the state under moral duty to avoid this war? As it seems, states have no reason to be bound by the constraints that were undertaken to advance their interests *ex ante* if doing so would *ex post* work against them.

The answer to this challenge is that although it is mutual benefit that underlies the moral validity of the contract that the Charter embodies, this validity is independent of the contract's success in promoting the parties' interests in each and every case. As emphasized in the previous chapter, the claims and correlative duties generated by contracts are valid even if *ex post* they benefit no one. These duties and claims are rigid—they do not lose their normative force in cases where respecting them does not benefit one of the parties.

3.3 Fairness: Minimally Just Symmetrical Anarchy

To be morally valid, contracts must be freely accepted. Presumably, they are freely accepted only if they are mutually beneficial. But *Mutual Benefit* is not all that matters. We assume that the contract that the Charter embodies ought to be fair too. States have the moral power to waive their right to wage pre-contractually just wars only if by subjecting themselves to the terms of the Charter, they do not cause or solidify injustice. That is, by entering the contract they do not create inequalities in the distribution of morally relevant goods like power, opportunities, resources, etc.

Does the *ad bellum* agreement pass this test? We assumed that states agree to avoid pre-contractually just wars because most future conflicts have a positive bargaining range. Alas, by its very definition, the bargaining range of a conflict is determined by might rather than by right; that is, by the probability that one of the parties will win the war and by the costs of the war to each of the parties. Both factors are mainly a function of military power; the peaceful resolution that the Charter favours over fighting will reflect the power of states rather than the implementation of justice between them. The worry, then, is that even if a contract that condemns justice-implementing wars is expected to benefit all parties—that is, even if the peaceful resolution that will be reached under this regime is Pareto-superior to war—the contract might still be unfair. Put more generally, the worry is that the contractarian approach stabilizes the status quo regarding power relations. How, then, can states be held to such a contract?

Our answer is that under circumstances of minimally just anarchy, a contract that condemns justice-implementing uses of force does not maintain or create unjust inequalities, but rather promotes justice. As the name suggests, circumstances are minimally just if they satisfy some minimalist criteria of justice. Following most theories of justice, we shall assume that facts about differences in welfare, power, resources, or capabilities between states and between individuals determine how just certain circumstances are.[15] Circumstances might be minimally just by virtue of three main features: (1) the unfair differences are not that significant, (2) the historical process that created them is not clearly unfair, and (3) the better-off individuals or states are not culpable for the unfair inequalities from which they benefit.

Minimal justice, in the sense defined above, has an important consequence. It creates normative uncertainty.[16] Had violations of international justice been (typically) grossly unjust, the relevant players could easily see that eliminating such violations by the means of war is justified overall. But it is not. In most cases, the normative facts—the knowledge of which is necessary in order to apply principles of pre-contractual morality that govern resort to pre-contractually just wars—are hard to obtain. Hence, in a regime governed by pre-contractual morality, states would not be able to determine with enough certainty whether or not preventive and subsistence wars were justified in general. Under minimally just circumstances, states ought to avoid most pre-contractually just wars because of the uncertainties that they would confront in deliberating whether to fight them. Indeed, states would have to avoid such wars even if they were totally impartial.[17] We argue in the following paragraphs that *Fairness* is satisfied because the Charter, which prohibits pre-contractually just wars, typically prohibits wars that should not be fought due to the normative uncertainties states confront under minimally just anarchy.

We develop this argument in three stages: (1) we present the uncertainties that states confront under minimally just circumstances; (2) we explain how these uncertainties should affect states' behaviour; and (3) we argue that most (but not all) wars that are prevented thanks to the Charter would not be fought by impartial states and suggest that this feature of the Charter makes it fair.

---

[15] See, for instance, Ronald Dworkin, 'What Is Equality, Part 1: Equality of Welfare', *Philosophy and Public Affairs* 10 (1981), 185–246; Gerald A. Cohen, 'On the Currency of Egalitarian Justice', *Ethics* 99 (1989), 906–44; Richard J. Arneson, 'Equality and Equality of Opportunity for Welfare', *Philosophical Studies* 55 (1989), 77–93; and his 'Luck-Egalitarianism and Prioritarianism', *Ethics* 110 (2000), 339–49; Larry Temkin, 'Inequality: A Complex, Individualistic, and Comparative Notion', *Philosophical Issues* 11 (2001), 327–52; and his 'Egalitarianism Defended', *Ethics* 113 (2003), 764–82.

[16] Here, we follow Andrew Sepielli, *Along an Imperfect Lighted Path: Practical Rationality and Normative Uncertainty* (PhD Dissertation, Rutgers University, 2010) 66. Let us introduce Sepielli's position through one of his examples. Suppose that the degree to which you believe in extreme 'meat-is-murder' vegetarianism is quite small (0.3). It turns out, however, that meals with meat are, on the whole, only slightly tastier than their vegetarian alternatives. What you have most reason to do is to avoid eating meat. If, however unlikely, eating meat is murder, then doing so is deeply wrong, whereas, if not, what you will have missed is merely the experience of eating meat. You should therefore behave as if you believed in extreme vegetarianism.

[17] See Richard Norman, 'War, Humanitarian Intervention and Human Rights', in Richard Sorabji and David Rodin (eds.), *The Ethics of War—Shared Problems in Different Traditions* (Aldershot: Ashgate Publishing, 2005), 204–6, arguing that the use of force by individual states to prevent violations of human rights would necessarily be selective and inconsistent, hence the need for an agreed-upon international body to make such decisions.

(1) A war whose positive effects outweigh its negative effects might be overall impermissible merely because of the means by which the positive effects are achieved. The exact scope of the means principle, according to which it is sometimes wrong to harm a person even as a means to pursue a great good, is largely indeterminate. Even if, say, the distribution of resources between states is clearly unjust, the amount of death and destruction caused by wars gives impartial states a basis for reasonable doubt as to whether a war whose goal is to reform this unjust distribution is justified overall.

Consider more specific uncertainties that states under minimally just circumstances confront. In the real world, billions of people are starving or on the brink of starvation, while a few hundred million are well off. Whatever one's theory of distributive justice is, this is clearly wrong. As it stands, however, this fact about the injustice of the global distribution of wealth does not entail anything decisive about the morality of using force, because we cannot ascertain who is culpable for the unfair distribution of resources.

To be sure, in certain cases the states and non-state actors responsible for the injustice are identifiable; yet even in these cases, it is hard to know whether a pre-contractually just cause for war exists. Suppose you are a citizen of Poor. In negotiating the borders of your country, generations before you were born, the founding fathers forfeited Water-Land, an area rich with perennial water springs. In their water-rich world, forfeiting Water-Land was dictated by prudence. The founding fathers of Poor could not have predicted how valuable Water-Land would become generations later. Of course, water has always been an essential resource; but the founding fathers had other water resources to rely on, resources which disappeared over time.

Does the old deal block Poor's claim to Water-Land? On the one hand, states have the authority to enter into agreements that redistribute the rights of the individuals that they represent; therefore, since Poor forfeited Water-Land for the sake of its citizens, the forfeiture is final and Poor has no valid claim to it any more. On the other hand, since the current citizens of Poor would never have authorized its founding fathers to alienate their rights, the founding fathers had no legal or moral right to do so. The current circumstances are so different that the contracting parties could not have intended to address them.[18] Hence, even when the agent responsible for the unjust poverty is identifiable, it is uncertain whether Poor has a claim against Wealthy for the water in its territory.

Preventive wars raise similar difficulties. They are morally superior to ordinary wars of national defence if fighting early is less costly to both sides. However, to determine whether a future war is a war of just prevention, states must assess whether or not a change in the balance of power is an encroaching unjust threat.

---

[18] This is based on Allan Gibbard, 'Natural Property Rights', *Noûs* 10 (1976), 77–86.

Their judgement must appeal to indications of intent and to political tendencies of the relevant state. That is, 'characterizing [changes in the balance of power] as threats is to characterize them in a moral way'.[19] Now, what makes circumstances minimally just is, among other factors, the complex institutional structure of states. Attributing intentions to such complex entities involves speculation and guesswork. True, intentions of political leaders can be perfectly perspicuous. Still, the checks and balances that a decent state imposes on any important political decision-making render the intentions of particular leaders less relevant. Therefore, cases in which the aggressive intentions of states are transparent are rare. The typical state of affairs is one of normative uncertainty ('normative' in the sense specified in the above citation from Walzer; assessing intentions or behaviours as *threats* assumes moral judgement).

(2) As indicated above, the normative uncertainty that follows from minimal justice should cause states to avoid most pre-contractually just wars. Arguably, justice typically requires bargaining rather than fighting, because of the normative uncertainty with respect to the exact scope of the means principle and with respect to the justice of the cause of the war. These uncertainties, on the one hand, and the certainty with respect to the huge moral cost of all wars, on the other, would bring states to avoid most wars under minimally just circumstances.[20]

(3) The Charter is fair because it promotes justice in the following sense: thanks to the regime that the Charter imposes, the wars states avoid are wars that they should avoid anyway, due to the uncertainties they confront. Typically, then, the war agreement allows partial states to overcome their partiality. States create a mechanism that prevents some Pareto-inferior wars. And luckily, most wars prevented by this mechanism would be unjustified anyway. Now, decently partial states have no interest in being more impartial than they actually are, and they do not sign the Charter in order to promote *justice*. They sign it to promote their narrow self-interest. Still, the system that emerges from this exchange of rights is preferable in terms of justice, compared to leaving states under the authority of pre-contractual morality. While in rare cases it bans wars that should have been fought in order to promote justice, in most cases it prevents wars that states should have avoided independently of the contract.

It follows that we answer the two crucial questions about *Fairness* in the negative: (1) Does the arrangement under assessment *maintain* the unfair inequalities in the actual state of affairs? Mostly, no—most wars banned under

---

[19] Walzer, *Just and Unjust Wars*, 79. Churchill's recommended policy in 1936 comprises a good example: 'he insisted that Britain would maneuver against Germany rather than France, on the ground that France, although apparently the strongest power on the continent had no aggressive intentions, while Germany was possessed by a will to dominate' (Marshall Cohen, 'Moral Skepticism and International Relation', in Charles R. Beitz, Marshall Cohen, Thomas Scanlon, and A. John Simmons (eds.), *International Ethics* (Princeton, NJ: Princeton University Press, 1985), 19.

[20] We follow here Sepielli, *Along an Imperfect Lighted Path*, 66.

the agreement are ones that should not have been fought anyway. (2) Does the arrangement *create* unfair inequalities of its own? No; the agreement is general and cannot be known in advance to discriminate against any party on the basis of the party's strength, prosperity, national identity, etc. We discuss one exception— the discrimination against stateless nations—in Section 3.6.

This is then the core argument for the fairness of the Charter under conditions of minimally just symmetrical anarchy. Contrary to the objection with which this section began, contractarianism is not conservative in the sense of maintaining or of strengthening existing power relations. It is, however, conservative in a different sense; namely, with respect to using force whose aim is converting minimally just circumstances into moderately or maximally just circumstances. Given the uncertainties involved in minimally just circumstances, use of force is morally risky; we ought to improve global justice by other means.

Let us discuss another natural objection to this move. One might argue that clear injustices constitute a just cause for war whatever the terms of the Charter may be. Consider the following scenario. A burdened society C(olonized)-Poor, which suffers from the greed of a rich superpower that exploits its citizens and its land, is a victim of manifest distributive injustice—especially if the rich state colonized C-Poor in recent history. It seems that C-Poor would be allowed to use force in order to free itself from the distributive injustice imposed on it even if the colonizer does not violate its territorial integrity. In other words, by consenting to the UN Charter, C-Poor would perpetuate the status quo that systematically discriminates against it; hence, such consent would not be morally valid. As clarified above, a regime that violates *Fairness* is morally mute.

This objection misses an important feature of the rules that states accept. Other things being equal, robust norms have a higher level of compliance compared to more subtle and complex norms. Had a proviso that explicitly permitted subsistence wars been an element of the Charter's prohibitive *jus ad bellum*, it would have been open to too many interpretations. As a result, partial states would appeal to the interpretation that best fits their narrow interests. Therefore, it is the shared interest of the contracting parties that the treaty-based *jus ad bellum* consist of robust, generic rules. This is why the Charter condemns *all* justice-implementing violence. Since these are mutually beneficial and fair rules, it is presumably true that states freely accept them. Their acceptance of the contract amounts to waiving the right to wage all pre-contractually just wars, even those subsistence wars that are clearly just.

Another important objection (to which we already hinted above) reads as follows. Wars of national defence might be fought against states that illegally pursue pre-contractual justice by the means of war. If there is uncertainty over whether it is justified to fight pre-contractually just wars, there must be similar uncertainty over whether the defensive war conducted in response is justified.

Our analysis allows for a simple answer to this concern. While states under minimally just circumstances suffer from uncertainty about principles of pre-contractual justice, much less uncertainty exists with respect to the terms that the Charter embodies and their appropriate application. When the contract is in place, the aggressiveness of states that violate it by initiating (perhaps pre-contractually just) preventive or subsistence wars is clear: they impose an imminent threat to the territorial integrity of the victim state. Consequently, there is relatively much less doubt that the victim state has a just cause for war, under the contract.[21]

To recapitulate, minimal justice generates normative uncertainty with respect to the justice of most subsistence and preventive wars, as well as other pre-contractually just wars. Most (imaginary) impartial states will not fight pre-contractually just wars, whereas partial states are likely to engage in such wars mainly because they are susceptible to errors in applying their pre-contractual rights. Instituting a regime that prevents some justice-implementing wars reduces injustice under the normative uncertainty from which states are likely to suffer. Therefore, by accepting the terms of the Charter, states bring about an outcome in which the overall violation of justice is reduced.

To use the Razian notion of legitimate authority, our argument entails that the *ad bellum* rule that prohibits use of force has legitimate authority over states in most cases (more accurately, the *legislator* of this rule is a legitimate authority). This is because, typically, use of force is impermissible independently of the contract that explicitly prohibits it. We stress, though, that the contractual duty exists even when a state can know *ex ante* that a pre-contractually just war is in its narrow interest and does promote justice. States undertake this duty by entering the arrangement that categorically prohibits pre-contractually just war.

We've been referring to states placing themselves under an arrangement that distributes rights and duties with regard to the waging of war. This does not necessarily mean an explicit act of commitment on their part. In most social contexts, the relinquishment of rights that constitutes the kind of contract we have in mind is less conscious and less explicit than in the case of boxers, for instance, entering the ring. Since the rules are mutually beneficial, such a relinquishment can nonetheless be ascribed to the relevant players. Contractarianism emphasizes the actual practice of following certain rules, as opposed to an express consent to them. This focus clearly applies to the basic *ad bellum* idea, namely, the unique value the UN Charter attaches to territorial integrity. Since the dawn of human history, the invasion of one's perceived territory—be it a family home, a tribe's *de facto* area of control, or a state's *de jure* political borders—has been regarded as the most obvious *casus belli*.

Insofar as international conventions are concerned, it is typically the case that the contracts we have in mind *do* originate in an explicit agreement between states

---

[21] Admittedly, in one respect, the justice of wars of national defence might be as doubtful as the justice of wars whose cause is pre-contractually just. Wars of national defence might well be disproportionate because of the harm they inflict on innocents. Moreover, due to the vagueness of the notion of proportionality, alongside objective difficulties in assessing future casualties, partial actors would tend to judge such wars as proportionate nevertheless.

that was expressly stated in the UN Charter. Still, such explicit undertaking of the rules concerning war can be ascribed only to states, not to their individual citizens, be they combatants or noncombatants. If the *ad bellum* agreement is fair and mutually beneficial, then as we have suggested in Chapter 2, its acceptance by states is morally effective: by accepting it, states consent to be governed by it, thereby waiving the moral right to use force in pursuing pre-contractual justice. Moreover, they allow the states against which they might aggress to protect that contractual right by force.

Does it obligate individuals? It might be worried that individuals do not accept the war agreement in the same way that they accept domestic law and therefore, this agreement has no effect on the distribution of rights among individuals. However, contractarianism denies a gap between national and international law. The denial follows from an assumption that seems to be part of the contractarian framework: by freely accepting a legal system, citizens authorize the state that enforces it to act on their behalf in the international realm. Hence, if fair and mutually beneficial, a treaty-based international law is as morally effective as the domestic law by which these individuals are governed.

Put somewhat differently, while the actual war agreement is signed by states, it is morally effective only if self-interested yet decent *individuals* would have signed it themselves. The war agreement is morally effective only if in signing it, states act on behalf of their citizens, such that by respecting their obligations under the treaty states make individual citizens safer, with better-protected rights. Of course, citizens' authorization of the war agreement is not explicit. Therefore, just as with other social rules, it is presumably true that they freely accepted the rules that they authorize their states to draft, if this set of rules satisfies the conditions of *Mutual Benefit*, *Fairness*, and *Actuality*. We argue, in other words, that citizens' widespread adherence to the legal system posited by the state is a manifestation of authorization; furthermore, if people would agree to these rules regardless—since they are mutually beneficial and fair—the rules are morally effective. (More on this below.)

3.4 A Hypothetical: A Story about Divided and Well-Ordered

To see how broad and pervasive the scope of the prohibition on first use of force is according to contractarianism, we apply it to the following hypothetical case.[22,23] Consider a state called 'Divided', which is comprised of four powerful ethnic groups. Members of each powerful group share different ethnic origins, societal cultures, and religions. The hostility between the groups is an essential element of

---

[22] This Section draws on Yitzhak Benbaji, 'State Self-Defense', in Larry May et al. (eds.), *Cambridge Handbook of Just War* (Cambridge: Cambridge University Press, 2018), 59–80.

[23] This case was inspired by the 1982 war Israel fought in Lebanon.

the history, the historical memories, and self-identification of their members. Unlike other political communities, these four ethnic groups do not form a united 'group agent' and, given the animosity and mistrust between their members, they cannot become such an agent. Furthermore, the weakness of the civil society in Divided makes it extremely vulnerable to external forces. After a long and bloody civil war, an unstable, compromise between the four big groups was achieved, each group controlling a proper subset of the state-institutions. Still, Divided is especially corrupt: the public interest plays no role in guiding the institutional decision-making and actions of its leaders. The main goal of role-holders in these institutions is to gain more political power for themselves within their groups and more political clout for the group to which they belong. As a result, mistrust, bribery, and other types of corruption become prevalent.

The other hero of our story is 'Well-Ordered'. Its regime is just and it is the national home of a civil society whose members share a societal culture. Well-Ordered requires some of Divided's territory for security reasons: Divided is unstable, and hostile groups within it permanently pose threats to its neighbours. Well-Ordered can now conquer the territory by using force. This would violate the sovereignty and territorial integrity of Divided. Still, since Well-Ordered is so much stronger than Divided, the invasion is expected to be bloodless. As members of the weakened groups in Divided, inhabitants of the conquered territory are under-represented by its regime. The occupation will not worsen their condition, nor would it compromise the fulfilment of their political rights.

The international law as we interpret it would not condemn changing the status quo in Divided by non-violent coercive measures. Well-Ordered might try to impose economic sanctions on Divided. It might campaign for secession, using propaganda to convince the residents of the territory that it seeks to annex to support the secession. Well-Ordered might openly or secretly support one of the groups in Divided, if this could further its security. Yet, according to the contractarian analysis, Well-Ordered ought not to change the status quo by using force. The war against Divided would be a crime against peace.

Compare the way the *jus ad bellum* contract treats the use of force to the way such a conflict is handled within domestic societies. I grow sunflowers in my yard. You put up a garage in yours, thereby depriving me of light. My sunflowers die. You harm me but do not wrong me.[24] You would wrong me if you brought about the same outcome by violating my sovereignty; if you ruined my sunflowers by trespassing in my yard or by using force against me.[25] Likewise, you might permissibly harm me by opening a competing business that lures away my

---

[24] Yet, in some jurisdictions, the zoning laws are such that even if you have a legal right to build a garage on your property, neighbours whose garden vegetables would get less sunlight as a result would have a cause to object to your garage-building and may even succeed in overriding your right.

[25] Ripstein, *Force and Freedom*, 78.

customers, but you ought not to bring about the same outcome by stealing a patent and selling it in the free market, even if both actions are equally harmful to me.[26]

The *jus ad bellum* contract, as we structured it above, regulates the means that a state might use in pursuing ends, rather than the ends being pursued. Use of force is an inappropriate means for promoting the good, even if it involves no bloodshed and has no bad effects. This seems to accord with our considered moral judgments: even those who believe that international law should allow humanitarian intervention against a brutal and oppressive regime concede that in less extreme cases, where a state is merely corrupt, dysfunctional, irrecoverably divided, etc., bloodless military intervention whose aim is annexation or a regime change is morally impermissible. True, it would be better if Divided ceased to exist and a different state were founded in its stead. Moreover, in terms of the contractarian argument, Divided has no claim-right against others' taking coercive measures to demolish its regime and/or redraw its borders. The point of the *jus ad bellum* contract is that these outcomes should not be brought about by the use of force, even if they cannot be brought about by non-military coercion.

The contractarian resolution of the Well-Ordered/Divided hypothetical reveals another crucial feature of contractarianism. It shows that contractarianism rejects Walzer's collective rights-based theory of the right against aggression. Walzer's core idea is that the right to political independence depends 'upon the reality of the common life it protects and the extent to which the sacrifices required by that protection are willingly accepted and thought worthwhile'.[27] Thus, in Walzer's view, the use of force against a state that supports, protects, and maintains a common way of life, a particular conception of justice, or group agency, is a violation of this state's right to political independence.

Contrary to this normative picture, according to contractarianism, Divided's right against aggression is not necessarily tied to the values that underlie the moral standing of states. Divided's right to defend itself by force from outside intervention is independent of these values. It has a right that no change in its legal standing will be brought about by use of force. Put generally, even if it is morally permissible to demolish a dysfunctional or clearly unjust state, doing so by force is ruled out by the *ad bellum* contract and entitles the victims to go to war to defend themselves.

What follows is a surprising version of *Moral Equality* at the level of *states*. On the basis of the considerations developed earlier, states agree that (significant) threats to territorial integrity comprise the only *casus belli*. The right to wage war does not depend on the war being instrumental in protecting values such as a collective way of life or on the moral character of the regime. Put simply, under contractarianism,

---

[26] Ripstein, *Force and Freedom*, 22.   [27] Walzer, *Just and Unjust Wars*, 54.

(almost) all states have a right to go to war in order to block threats to their territorial integrity (provided that doing so is a last resort and is not disproportionate in terms of the expected harm to civilians). The rationale of Article 51's 'inherent right to self defense' in Divided-like cases is to deter decent states from engaging in wars whose aim is the implementation of pre-contractual justice.

3.5 Wars of Independence

As we interpreted it above, the war agreement regulates disputes between decent states. The *ad bellum* agreement applies to disputes that result from the 'inconveniences of the state of nature'; namely, disputes over territories, natural resources, contractual rights, etc. Decent states that undertake the Article 2(4) duty not to pursue pre-contractual justice by means of war fulfil their duty to improve protection of their citizens' pre-contractual rights to life, bodily integrity, property, etc. The war agreement is mutually beneficial and fair, especially because the pre-contractual rights of individuals are better protected thanks to the arrangement that underlies the agreement.

Yet, in its current form, the war agreement systematically discriminates against non-state actors in general, and stateless nations in particular. To see why, note that the war agreement protects not only the classical pre-contractual rights—to life, bodily integrity, etc.—but also the pre-contractual right of individuals to be full members in a self-determined political society. Self-determined political societies are important since they maintain the way of life and the national/cultural identity of their members. And by 'sanctifying' the territorial integrity of states, the war agreement usually protects the right to self-determination of nations who already have a state. (As we noted in the previous section through the case of Divided, rarely does a state have no right to its political independence despite the fact that aggression against it is a crime against peace.) Thus, stateless nations are disadvantaged by the territorial definition of aggression. With no state of their own, they have no borders whose crossing could serve as a *casus belli* under the contractarian approach. Moreover, in some cases, they have a pre-contractual right to use proportionate force in order to realize their right to self-determination. But under the war agreement as we have so far construed it, this right is denied. States are advantaged by this prohibition, but one might suspect that stateless nations attain no parallel advantage from it.

In response to this challenge, we suggest interpreting the war agreement such that the parties to the *ad bellum* contract are not only states, but stateless nations as well.[28] Non-state actors who have ambitions to form states of their own should

---

[28] States—especially those in conflict with national groups seeking independence—tend to portray all such groups as 'terrorist', or as connected to 'global terrorism'. But as Michael Gross shows in *The*

be viewed as bound by the war agreement.[29] This is simply because a legal system should represent anyone who is governed by it, and should take into account the interests of anyone who might be affected by it. As we shall see shortly, this move immediately implies that at times, wars of national independence might be permissible. In this section, we explore the just cause for a war of independence, according to contractarianism.

As a starting point, we suggest that decent states are related to decent stateless nations just as decent states are related to each other. Therefore, the agreement that regulates use of force assumes that stateless nations confront minimally just symmetrical anarchy. Hence, in most conflicts between states and stateless nations, a bargaining range exists. The huge costs involved in war and the uncertainty regarding its success imply that, even in conflicts between states and stateless nations, bargaining is typically *ex ante* better for both sides than fighting. Under the contract, the contractual right of decent stateless nations to use force against decent states is extremely restricted. They should carry out their struggle for self-determination by bargaining rather than fighting.

Moreover, most disputes about national liberations involve normative uncertainty. Whether or not collectives have a right to their own state is far more controversial and indeterminate than is often assumed. Not every ethnic, national, or religious group has an automatic right to political self-determination. Even when a stateless nation is entitled to political independence, this claim only comprises a *prima facie* one that often must compete with other moral considerations, such as the rights of the current inhabitants to the required land.[30] Even among groups that have a right to self-determination, realization of the entitlement to political independence need not take the form of a full-fledged state; greater autonomy within the state in which they are currently residing may be sufficient. In sum, whether or not a national group is entitled to a sovereign state is typically an open question.

If the assumption that stateless nations confront minimally just symmetrical anarchy is true, the expected harm of a war of independence is so great that typically, the parties have prudential reasons to compromise for less than full sovereignty, or for sovereignty over a smaller territory than they think they deserve, instead of resorting to war. Usually, the death and destruction that inevitably accompany acts of war are too severe in comparison with the expected

---

Ethics of Insurgency: A Critical Guide to Just Guerrilla Warfare (Cambridge: Cambridge University Press, 2015), this tendency is unwarranted. While many guerrilla organizations resort to terrorism, 'most are neither terror organizations nor a party to global terrorism' (2).

[29] See George P. Fletcher, *Romantics at War: Glory and Guilt in the Age of Terrorism* (Princeton, NJ: Princeton University Press, 2002), 6, noting that 'the Civil War was a war in the fullest sense even though no one recognized the Confederacy as a state'.

[30] A good survey of the different considerations can be found in Avishai Margalit and Joseph Raz, 'National Self-Determination', *The Journal of Philosophy* 87 (1990), 439–61.

good that stateless nations seek to realize. Hence, stateless groups have an interest in undertaking a regime under which independent wars would be ruled out even in the rare cases in which it is in their best interest to fight. Since a normative mistake (as to whether or not the right of political self-determination exists) is quite likely under such circumstances, individuals and groups have difficulties in overcoming biases when they reason about national liberation. That is, the legal prohibition on the use of force is not only mutually beneficial, it is also fair as it typically bans wars that should not be fought, independently of the contract.

What, then, should be considered a just cause for a war of independence against a decent (yet partial) state? When does a national group have a right to rebel for the sake of political independence, if a decent state denies it this right? Wars of independence have a just cause when a (justly or unjustly) occupying power represses by force the (just or unjust) national aspirations of a national group. Fighting back under such circumstances is permitted even if the group in question has no right to an independent state.

This interpretation is almost a mechanical extension of the war agreement between states. Under the nonstandard interpretation offered here, first use of force against a group seeking political independence is a crime against peace. It is like the violation of territorial integrity in wars between states. The violation of the prohibition on the use of force is relatively discernible, and the parties to the contract aim to prevent use of force by deterring the parties from resorting to war. They agree to confer a contractual right to use defensive force in pursuing this end.

If this interpretation of the war agreement is correct, the suspicion that it systematically discriminates against stateless nations is groundless. The just cause of a war of independence is less strict than the just cause of defensive wars. Violent oppression is sufficient; if members of stateless nations are citizens of the oppressing state, then harsh police actions (which might be legitimate in fighting criminal gangs) comprise a just cause for a war of independence. The question whether fighters were authorized to fight in the name of a national group is important for assessing the justice of the war. This issue would be debated between militants and the state fighting them.[31] In many cases, however, it is quite clear that an NSA represents a national community, and that a state exercises force against militants in this organization. In these clear cases, the war agreement allows wars of independence.

As we understand it, the war agreement is egalitarian in another important sense: even those who fight unjust wars of independence are entitled to the protection the Laws of Armed Conflict confer on enemy soldiers. We will discuss this issue in greater detail in Chapter 6.

---

[31] For a detailed discussion, see, Benbaji, 'Legitimate Authority in War'.

Our analysis entails that there may be national groups that would be left with no effective way of realizing their right to political self-determination, just as there may be poor states that have no effective way to remedy the distributive or corrective injustices from which they suffer. The fact that under a regime that prohibits the use of force, some injustices that can be eliminated by war are left unaddressed has been acknowledged already with respect to subsistence wars. Although in rare cases subsistence wars might be pre-contractually justified, the risk of error is very high. Hence, with the exception of indecent states that rob the resources of poor nations and repress them, it is better for the poor countries to subject themselves to a rule that prohibits subsistence wars in all cases. In exceptional cases, such a rule might cause (or solidify) unjust inequalities, but for the most part it will reduce injustice; namely, wars that ought not to have been fought in the first place.

3.6 Self-Help vs Global Police or Global Court

One might think that despite the potential of a self-help regime to enforce the prohibition on aggression by permitting defensive wars, a serious concern for human rights should lead to other arrangements. The self-help based regime that the Charter institutes ought to be replaced by some other arrangement. The purpose of this section is to explore these alternatives and show that they are morally inferior to the one proposed above, mainly because of the way they compromise on the ideal of fair political participation.[32]

The first alternative is the one Rodin envisions; namely, an international community governed by an ultra-minimal, universal state.[33] A state is ultra-minimal if it consists 'solely in the establishment of a world monopoly of military force together with a minimal judicial mechanism for the resolution of international and internal disputes'.[34] This global state is to be distinguished from the more 'substantive conception of world government with a broader administrative mandate' considered and rejected by political theorists from Kant to Rawls.[35] An ultra-minimal global state would render armies redundant; violent confrontations between states—just as those between individuals—would be criminal events, to be addressed and contained by the police of the global state.

Unlike Rodin, Jeff McMahan presumes that under the current circumstances, national armies are the means by which the moral prohibition on aggression is

---

[32] This Section relies on Yitzhak Benbaji, 'Against a Cosmopolitan Institutionalization of Just War', in Yitzhak Benbaji and Naomi Sussmann (eds.), *Reading Walzer: Sovereignty, Culture and Justice* (Routledge, 2013), 233–56.
[33] David Rodin, *War and Self-Defense*, chapter 7.   [34] Ibid., 184–5.   [35] Ibid.

most efficiently enforced.³⁶ The current international legal regime is defective not because it lacks a transnational monopoly on legitimate force, but because it allows states to determine whether their claims are just and whether war is the right way to enforce them. The obvious downside of an arrangement that authorizes states to determine the justice of their cause for war is that states are partial, and their political leaders may be dishonest or even corrupt and vicious. Hence, they may utilize claims of self-defence as a pretext for their illegitimate interests.

Thus, McMahan offers a transnational court-like institution entrusted with the prerogative to rule as to whether a war is just. The court will facilitate the refusal by soldiers to fight in unjust wars by securing their free access to the relevant information. If the court declares a war in progress to be unjust or illegal, and if this judgment is widely disseminated, that will weaken the excuses available to those fighting in it. Their ability to plead non-culpable ignorance would diminish.³⁷

Now, we will assume (though McMahan tries to avoid this assumption) that under this regime, the court can issue legally binding instructions not to go to war, or to halt the war. Such powers openly limit the scope of the inherent right of self-defence conferred on states by Article 51 of the UN Charter. Under the new regime, states have no legal power to determine whether they are under aggressive threat. An authoritative court must license a war. Importantly, the imagined regime undermines the formal sovereignty of states by denying them a legitimate way to exit from the jurisdiction of the court. The denial of this right is crucial because once states are declared as aggressive, they are likely to do whatever they can to undermine the court's *de facto* authority over them. Thus, the law must prohibit the exit of aggressive states in order to secure court authority in cases where it is needed.

One might consider such regime changes unfeasible; but we take a different path. We question the assumption that the self-help regime that governs the decentralized society of states is, all things considered, inferior to the centralized regimes that Rodin and McMahan envision. States, we submit, have a weighty moral reason against 'ceding powers formally to transnational agencies, even as they continue to engage in forms of *de facto* cooperation that have functionally equivalent ffects'.³⁸ Put differently, the transnational institutional scheme to which states ought to subject themselves in order to fulfil their fundamental duties should be consent-based. At least *prima facie*, the rules of global politics to which they subject themselves should not compromise the *de jure* sovereignty of states.

---

³⁶ Jeff McMahan, 'The Prevention of Unjust Wars', in Benbaji and Sussmann (eds.) *Reading Walzer* (Abingdon: Routledge, 2014), 233–55.
³⁷ Ibid., 242 ff.
³⁸ See Alexander Wendt, 'A Comment on Held's Cosmopolitanism', in Ian Shapiro and Casiano Hacker-Cordon (eds.), *Democracy's Edges* (Cambridge: Cambridge University Press, 1999), 128.

This is so for the following reason. A decentralized society of states better realizes the ideal of fair political participation—an ideal that reflects the concept of politics as a collective enterprise whose purpose is to regulate the lives of those affected by it. Since self-control is important to us, participation is required by virtue of the centrality of politics in our lives. Participatory politics are politics in which all relevant parties have an equal say in crucial questions such as who will exercise political power over them and how such power should be wielded.

Now, the entitlement of individuals to partake in the decision-making process of their group derives from the fact that they are expected to abide by the decisions made even if they disagree. This logic strongly suggests that the participatory ideal does not entitle individuals to participate in political deliberations that will not affect their lives and wellbeing. Moreover, the degree to which people participate in the formulation of the regulations to which they are subject tends to be negatively related to the degree to which the relevant regulative institution is responsive to external influence (i.e. the influence of those who will not actually be subjected to those regulations). Including outsiders in a group's decision-making process thus strongly infringes the rights of members' participation therein, and in this sense is less fair. In other words, fair participation is compromised if outsiders—those whose interests and wellbeing are not affected by a certain policy— have influence on its direction.

The institutionalization of just wars, which Rodin and McMahan favour, has exactly this effect. Transnational institutions of the type they envision would make international politics less participatory compared to international politics formally governed by states and state institutions.

The ultra-minimal state would respond to the preferences of all its citizens, even in cases in which it resolves armed conflicts between two small and isolated national groups. The same is true of the *jus ad bellum* court envisioned by McMahan. A regime that prevents soldiers from participating in a war that some external supra-national court considers to be unjust is costly in terms of fair political participation. The transnational courts should be accountable not only to the peoples immediately affected by their rulings but also to outsiders who are often ignorant of the causes of war and uninterested in its consequences. Both regimes compromise fair participation by undermining the *de jure* sovereignty of states.

The decentralized regime we described in the previous sections is a bottom-up regime that excludes outsiders from political decision-making processes regarding regions and conflicts to which they are unrelated. It insists that the involved states, rather than transnational courts, ought to represent their citizens under circumstances in which international conflicts might affect their wellbeing. The relevant parties might institute an ad hoc court whose impartiality can be secured by mediators that both parties trust. They can agree in advance to accept its verdict.

But outsourcing authority (as it were) to a permanent international court would be unfair to insiders.

Now, we fully concede that this price of harming self-determination through political participation ought to be paid if indeed under the regimes envisioned by Rodin or McMahan, the world would be better in terms of human rights fulfilment. If, thanks to the minimal state and/or the international court, there would be less large-scaled violent clashes between states, the ideal of political participation should be compromised. However, there is no reason to expect that this will be the happy outcome of these reforms. True, the call for a significant strengthening of international institutions and regulations is a natural response to the fact that the UN Charter allows for defensive wars that might be pre-contractually unjust. Arguably, the Rodin/McMahan regimes are more respectful of human rights, as they outlaw participation in unjust wars. But these proposed reforms in international institutions are unlikely to reduce the amount of violence for three reasons: (1) the new bodies they propose—be that a global government or an international court—will find it hard to get a real grip on the relevant factual and normative facts in armed conflicts around the globe; (2) the individuals composing of these institutions will have only limited success in transcending their own loyalties and prejudices; and (3) as a result, these bodies will suffer from a chronic lack of trust on the part of both politicians and laypeople.

3.7 Conclusion

This chapter offered a contractarian answer to two main questions regarding the UN Charter. First, why aren't just distribution, just prevention, and just regime change legitimate causes for war? Second, why do states have a right to protect their borders even if these borders are relatively morally insignificant?

We argued that against the background of a minimally just symmetrical anarchy, the Charter is a morally valid contract by which a system of contractual rights and duties is determined. States waive their right to launch preventive wars and subsistence wars and acquire a right to go to war in defence of their territorial integrity. States might be pre-contractually permitted to engage in wars that are contractually forbidden (like preventive wars or wars of regime change); they may be pre-contractually forbidden to launch wars that are contractually permitted (defence of mere territorial integrity). As argued in Chapter 2, when specified conditions are satisfied, legal systems might change the distribution of rights and duties in all fields of social activity.

The first purpose of the United Nations is:

> [t]o maintain international peace and security, and to that end: to take effective collective measures for the prevention and removal of threats to the peace, and

for the suppression of acts of aggression or other breaches of the peace, and to bring about by peaceful means, and in conformity with the principles of justice and international law, adjustment or settlement of international disputes or situations which might lead to a breach of the peace.

The second purpose is '[t]o develop friendly relations among nations based on respect for the principle of equal rights and self-determination of peoples'.

While many people assume that the right to national defence involves the second purpose (states' right to self-determination), we defend the view that it is really anchored in the second purpose (namely, the value of peace). The Charter is built on the assumption that global political and distributive injustice should be implemented by peaceful means, not by war. The right to national defence is designed to enable states to address aggression and to deter states from using force in resolving their disputes with unjust states or non-state actors.[39]

Note that the permissions and prohibitions entailed by pre-contractual morality are not 'deeper' in any significant sense of the term than those entailed by contractual morality. Once a contract is valid—regulating the arrest of suspects, as in POLICEMAN; the handling of money, as in BANKER; the behaviour of drivers, as in ACCIDENT; or the engagement in war, as discussed in this chapter—the relevant parties have the moral right to comply with it. That is precisely the point of contracts; that under most predictable circumstances, they are followed regardless of what would be required in the absence of a contract. We shall return to the assumed depth of pre-contractual morality later in the book.

The contractarian view of *jus ad bellum* implies an interpretation of the two main constraints on waging war; namely, last resort and proportionality. First, there is no just cause without last resort. When the former is satisfied—when the enemy crosses the border or is about to do it—using force to stop the invasion is usually the only resort available. As we shall see in the next chapter, proportionality is interpreted by reference to the limitations that states would undertake in enforcing the prohibition on first use of force by self-help.

Our interpretation implies that decent national groups struggling for political independence are also parties to the *ad bellum* contract. The war agreement assumes, however, that states and stateless nations are related to each other just as decent states are related to each other. If it is reasonable to presume that states confront minimally just symmetrical anarchy, it is just as reasonable to presume

---

[39] Our view is not far from that of Rawls's who assumes that states are under a duty of non-intervention, that they have a right to self-defence when their sovereignty is threatened and that they are parties to 'agreements that bind them'. Unlike Rawls, though, we deduce the right against intervention from the contract into which states ought to enter. See John Rawls, *The Law of Peoples* (Cambridge, MA: Harvard University Press, 1999), 37. We replaced 'peoples', which is the term used by Rawls, by 'states', which is what we think he has in mind here. For the problematic use of 'peoples' in this context, see Philip Pettit, 'Rawls's Peoples', in Rex Martin and David A. Reidy (eds.), *Rawls's Law of Peoples* (Oxford: Blackwell, 2006), 43.

that stateless nations struggling against states for self-determination confront such circumstances. Thus, these groups, too, waive their right to use force to pursue pre-contractual justice by means of war, unless they face a clear attempt to wipe out their national ambitions and demands by force.

Finally, as we show in subsequent chapters, under our interpretation, the *ad bellum* regime divides the moral labour. It is the role of political leaders (not of individual combatants or officers) to initiate just wars, even if these individuals justifiably believe that doing so is permissible and even mandated. The parties to the *ad bellum* contract agree that attacks initiated by non-politicians—be they combatants or noncombatants—are regarded as criminal and that their agents do not enjoy the protections and privileges of the war agreement. In traditional terminology, such attacks would violate the condition of legitimate authority.

The question of when wars may be started is closely related to the questions of which aims the warring parties should pursue and when such wars should come to an end. These topics are addressed in the next chapter.

4
The Aims of Just Wars and *Jus Ex Bello*

4.1 Introduction

In the previous chapter, we offered a contractarian argument for a legal system that confers on states a right against aggression. We defined aggression as a crime against peace of a particular form—an armed invasion into another state's territory (or an imminent threat to do so). We argued that the contract into which states in a minimally just symmetrical anarchy ought to enter categorically prohibits initial use of force, even, and in particular, as part of pre-contractually just wars. We further noted that under the war agreement, this contractual duty can be enforced by self-help only. The self-help clause is implemented through a right to go to defensive wars against states that use force in violation of the contract.

Defining and prohibiting aggression is one central element of the contractarian theory of *jus ad bellum*. But just war theory also determines the appropriate response to this crime. In this sense, just war theory resembles a penal code that not only defines and prohibits murder but also determines the punishment for it, and a civil code that not only defines private wrongs but also determines the appropriate compensation for them. So far, we have said almost nothing about the way crimes against peace—understood by contractarianism as violations of territorial integrity—ought to be addressed. We tackle this issue in the present chapter. Exploring the appropriate response to aggression involves inquiring about *the aims of defensive wars*. And such inquiry, in turn, implicates the question of when wars should come to an end. We ask, in other words, when wars can be said to have achieved their just aims. This chapter elaborates on the rules for the termination of war, viz., *jus terminatio*[1] or *jus ex bello*[2] (the expression we shall be using here).

In McMahan's view, the just aim of war and the just cause of war are two sides of the same coin. As he puts it, 'any engagement in war requires a just cause... [and]

---

[1] See David Rodin, 'Two Emerging Issues of Jus Post Bellum: War Termination and the Liability of Soldiers for Crimes of Aggression', in Jann Kleffner and Carsten Stahn (eds.), *Jus Post Bellum: Reflections on a Law of Transition from Conflict to Peace* (The Hague: T.M.C. Asser Press, 2008), 53–75.

[2] Darrel Moellendorf, 'Jus ex Bello', *Journal of Political Philosophy* 16 (2008), 123–36.

when the just cause of a war has been achieved, continuation of the war lacks justification and is therefore impermissible'.³ Against this view, we argue in this chapter that *jus ex bello* and *just cause* are distinct; that the war agreement draws a distinction between the rules that govern resort to war and the rules that govern its continuation, aims, and termination.

We proceed as follows. In Section 4.2, we contend that the reduction of just cause to just aims betrays the contractarian logic of the war agreement. In Section 4.3 we present the (distorted contractarian) logic underlying the idea of 'uncompromising wars' and offer a contractarian argument for a rule that categorically prohibits such wars. In Section 4.4 we argue for a rule according to which conventional acts like surrender force the termination of war on the victor. In Section 4.5 we develop two asymmetries between the justice of resort to war and the justice of its continuation. More specifically, we show that the conditions of proportionality and of probability of success function differently in constraining resort to war and in constraining its continuation.⁴

4.2 Just Cause vs Just Aims

The natural thought that McMahan illuminatingly explicates is that the just cause condition functions as a restriction on the type of *aim* or *end* that may legitimately be pursued by means of war. Since wars involve intentional and collateral killing, given the strong presumption against killing, the aim of a just war must be self-defence: 'only certain types of aims—such as self-defense against an unjust attack—can provide a justification for killing'.⁵ The just aim of war follows immediately—prevention of the wrong threatened or caused by the aggressor:

> There is just cause for war when one group of people...is morally responsible for action that threatens to wrong or has already wronged other people in certain ways, and that makes the perpetrators liable to military attack as a means of *preventing* the threatened wrong or *redressing or correcting* the wrong that has already been done.⁶

McMahan's way of understanding the relations between just cause, just aim, and *jus ex bello* might be thought to underlie the contractarian approach to the *ex bello* question. At face value, this understanding straightforwardly implies an answer to the question addressed in this chapter; namely, if the crime of aggression is the

---

³ Jeff McMahan, 'Just Cause for War', *Ethics and International Affairs* 19/3 (2005), 2.
⁴ This chapter revisits Daniel Statman, 'A Contractarian View of *Jus Ex Bello*', *Ethics* 125 (2015), 720–50.
⁵ McMahan, 'Just Cause for War', 4.   ⁶ Ibid., 8, italics added.

posing of an imminent threat to the territorial integrity of states, wars are to be terminated when they achieve the goal that justified them in the first place; viz., when they remove the aggressor's threat to sovereignty and territorial integrity. Ending war short of achieving this goal seems to undermine any justification the war might have had. Continuing it when this end has been achieved has no just cause.

Let us elaborate a bit further on this line of argument before rejecting it. Just war theory—as contractarianism structures it—assumes *as if* the society of states is governed by a 'just' arrangement; namely, an established, conventional, widely accepted arrangement, which makes for regional (or global) peace. Aggression comprises a violent disruption of this arrangement, and as such constitutes a violation of the territorial integrity of states, which might be repaired by a defensive war. Therefore, it might be thought that the only legitimate aim of defensive wars is to restore the peacefulness of the status quo ante; viz., to maintain the territorial integrity and the sovereignty of the victim of aggression. As we insisted, it is not a *just* peace but rather peace itself, as it existed before the aggression, which grounds the territorial rights of states. That is, the war agreement does not assume that the current distribution of territories among states is just; it merely prohibits using force in pursuing justice. One might naturally conclude that defensive just wars should stop once the status quo ante is restored.

On reflection, however, this way of deducing the just aims of war and the rules for its continuation runs against the contractarian logic underlying the war agreement. Arguably, according to contractarianism, in a symmetrical anarchy, wars against aggressors can and should have other aims in addition to eliminating the imminent threat to the territorial integrity of the defending state. After all, the society of states has no political centre that can deter states from aggressing against other states, and no political centre that can impose punishments for crimes against peace whose aim is the prevention of further crimes. The war agreement is about maintaining peace. As such, the agreement should devise legally permissible means for the defender to deter the aggressor (as well as other potential aggressors) from resorting to unjust wars. In other words, the rules that govern the just aims of war and the rules that govern the termination of war should enable defenders to deter their potential enemies. Additionally, since a preventive war might be Pareto superior to the defensive war it is likely to prevent (the case of an 'optimal preventive war'), the rules governing *jus ex bello* should allow the just victor to continue the war so as to prevent the aggressor from becoming involved in further aggressive wars.

The fact that contractually, just wars comprise a means to deter potential aggressors may suggest that a defender has the right to weaken the aggressor's army in a way that prevents the latter from being able to resume the war in the near future. This logic may even suggest that the just side might occupy some of the territories of the aggressor in order to force a peace treaty that makes further aggression less likely. Finally, and most radically, one might assume that deterrence and prevention legitimize what is known as 'uncompromising wars', whose aim is total destruction of the military forces of the enemy.

In the next section we explicate the logic of such wars and explain why it should be resisted. We argue that given the biases, stress, and self-righteousness of the warring parties, they have a good reason to agree on a rule that forbids uncompromising wars and interprets the permissible aims of war in a narrower way—which is still wider than that proposed by McMahan.

4.3 The Logic of Uncompromising War

The political science literature commonly distinguishes between military victory and political triumph. When countries go to war, defeat of the enemy army is not an end in itself but a means to some political end, which is distinct from the destruction or disablement of the enemy army. This end could be averting aggression, gaining security or strategic status, access to the sea, territory, economic benefits or any combination of these (or other) gains. Victory on the battlefield is not always sufficient to guarantee attainment of these political aims. If the cost of victory is too high, in terms of casualties or money, the victorious entity might emerge from the war in a worse situation than it would have been in had it refrained from fighting. Moreover, the ability to achieve the political aims of wars depends not only on objective facts, such as the occupation of certain territory, weakening the enemy army by killing soldiers, or eliminating military objects. In many cases, whether or not a war is successful depends on subjective perceptions; namely, on which side is *perceived* as the victor by the respective sides and by the international community. If the side defeated on the battlefield manages to create the impression that it fought heroically and prevented its enemy from achieving its aim, it might emerge as the real winner. Consequently, it might be in a better position in the post-war negotiations. Johnson and Tierney make this point well:

> In international relations, military victory, or indeed the gain of any tangible prize at all, is neither necessary nor sufficient for people to think a leader has won. Not necessary because victory can be obtained despite net losses; not sufficient because even substantial gains do not guarantee that people will view events as a success... Quite often, one side can exploit geography, technology, and strategy to defeat an opponent militarily, yet still emerge as the perceived loser, with all the tribulations that this status involves.[7]

---

[7] Dominic Johnson and Dominic Tierney, 'Essence of Victory: Winning and Losing International Crises' *Security Studies* 13 (2003–04), 350, cited by Mandel, 'Reassessing Victory in Warfare', 464.

How common is it for wars to end without the military victors reaping the fruits of their victory? In Robert Mandel's estimation, quite common indeed: 'after the end of the Cold War, regardless of the margin of victory, it has been rare for military triumphs in battle to yield substantial postwar payoffs'.[8] Moreover, the decline 'in the proportion of wars in which there is a clear-cut winner or loser' casts doubt on the notion that wars necessarily end in either victory or defeat.[9]

The failure to translate success on the battlefield into political accomplishment is nowhere more prevalent than in attempts to intervene in civil wars, which have constituted the majority of wars since the Second World War.[10] According to Daniel Byman and Taylor Seybolt, 'military interventions with the intention to bring about lasting peace in a violent communal conflict fail or even backfire far more often than they succeed'.[11] The significance of these observations cannot be overstated; many thousands of people killed and injured, enormous economic losses, destruction of infrastructure—all without being able to determine who the winners are and, even when the victors are agreed upon, without them reaping substantial payoffs. Indeed, this gives rise to the 'riddle of war': how can humans bring about such mayhem upon both their enemies and themselves, with so little gain? It also confirms the point made earlier about the rationality of an arrangement that *ex ante* prohibits national liberation wars.

What seems to follow from these considerations is that if a state believes that it is in its (narrow) interest to go to war, it should be willing to follow the war through to the end; namely, to the clear defeat of its enemy. This notion brings to mind Clausewitz's view about the aim—the 'essence'—of war and makes war seem logical and natural: 'The very concept of war will permit us to make the following unequivocal statements: (1) Destruction of the enemy forces is the overriding principle of war...(2) Such destruction of forces can usually be accomplished only by war.'[12] As James Whitman noted, some of Clausewitz's contemporaries conceded that annihilation was a valid aim, but only against 'savages', not against civilized Europeans. Clausewitz explicitly objected to this

---

[8] Mandel, 'Reassessing Victory in Warfare', 461.   [9] Ibid., 465–6.

[10] According to Lotta Harbom and Peter Wallansteen, 'Armed Conflict and Its International Dimensions, 1946–2004', *Journal of Peace Research* 42 (2005), 'a total of 228 armed conflicts have been recorded after World War II and 118 after the end of the Cold War. The vast majority of them have been fought within states' (623).

[11] See Daniel Byman and Taylor Seybolt, 'Humanitarian Intervention and Communal Civil Wars', *Security Studies* 13 (2003), 33. We owe the last two references to Kersti Larsdotter, 'Culture and Military Intervention', in Angstrom and Duyvesteyn (eds.), *Understanding Victory and Defeat in Contemporary War* (New York: Routledge, 2007), 206.

[12] See Carl von Clausewitz, *On War*, translated and edited by Michael Howard and Peter Paret (Princeton, NJ: Princeton University Press, 1976, originally published in 1832), 258, cited by Azar Gat, *The Origins of Military Thought: From the Enlightenment to Clausewitz* (New York: Oxford University Press, 1989), 206.

understanding, insisting that annihilation is the proper end of all wars, even against Europeans.[13] The claim is not a historical one (suggesting that all actual wars have involved annihilation), but a conceptual one regarding the basic logic of war, followed by a claim about the rationality of the self-interested party: how wars ought to be fought by such an agent.

This doctrine contradicts every aspect of traditional just war theory. Indeed, saying that the 'very concept of war' entails the 'principle' of annihilation might sound horrific to the current philosophical ear. And yet, our contractarian reading of *jus ad bellum* might lead to a related doctrine—that of uncompromising war. Under contractarianism, winning is never an end in itself. It is a means of defence from aggression, a way of undoing its harms and of deterring the actual and potential aggressors from violation of the Charter's prohibition on the use of force. In some cases, we conjecture, these aims cannot be secured if the military power that undertakes the aggression still exists. Such cases necessitate a complete removal of the ability to even begin hostilities, which means the destruction of the enemy's military capability.[14] A further contractarian consideration for conferring a right to total destruction of enemy military forces would analogize defensive wars to justified punishments in domestic societies. The law that entitles an institution to impose impediments is justified only if the punishment deters people from committing the relevant crime. Can the right to uncompromising wars be justified in the same way?

In an uncompromising war, the parties continue the struggle until clear and unambiguous defeat of the enemy army is achieved. They need not inflict more harm on the civilian population than 'compromising' wars do. In reality, however, they almost always do. If the enemy army is pursued and fought against relentlessly, that naturally increases the risk to civilians as well. Uncompromising wars also tend to impose greater destruction on civilian infrastructure—on buildings, roads, bridges, industry, agriculture, and so on.

If these aspects of uncompromising war seem unsettling, let us consider their underlying logic in greater detail. Consider a decent aggressor that erroneously believes that it has a good moral reason for going to war, despite the costs and the risks involved. The justification is pre-contractual; it is grounded in the goods that the aggressor hopes to reap (territory, resources, etc.), or in some ideology it holds (to halt the spread of communism, etc.). Whatever its exact nature may be, this reason provides the state with a powerful motivation to resort to a war that it strongly believes to be just. Insofar as this motivation exists, the victim of this aggression will not live in peace. Since it is almost never within the power of the

---

[13] See James Q. Whitman, *The Verdict of Battle: The Law of Victory and the Making of Modern War* (Cambridge, MA: Harvard University Press, 2012), 233.

[14] For the sake of the present discussion, we ignore the transformation in Clausewitz's thought in 1827 in which he also gave space to the notion of limited war. For this transformation and its meaning, see Gat, *The Origins of Military Thought*, 217.

threatened party to transform the fundamental motives of its enemy, what they can more realistically hope for is (a) to disable the military force of the enemy in a way that would make attacks in the near future much less feasible, and (b) to strike the enemy so badly that it will be deterred from initiating another violent round. Uncompromising war is not necessary to eliminate the threat in every case; yet, some decent aggressors might be so stubborn that only their total destruction could remove the threat they pose.

This logic seems to square nicely with Isabelle Duyvesteyn's view, expressed at the end of her concluding thoughts on victory and defeat:

> Without wanting to be accused of warmongering, we should not forget, as has been found time and again, that the most stable form of peace is achieved after a clear-cut military victory.[15]

A clear-cut victory is important also because (as noted above) what determines the post-war situation are not just objective events, but the perception of defeat in the minds of the warring parties and of the international community.[16] If the enemy retreats with heavy casualties but manages to turn apparent failure into a legend of heroism and sacrifice, the military defeat might fall short of deterring it from future aggression. Rich and diverse means exist to manipulate data. The just side needs an unambiguous victory, which will make it impossible for the defeated party to deny or reinterpret.[17] In those cases, Clausewitz's 'destruction of the enemy forces' with its unavoidable effects on civilian life might be necessary.

Notably, even harsh destruction of the enemy's military capacities might turn out to be insufficient to prevent the next war if the peace (or surrender) terms are not enforced with adequate determination; this was the case with the Versailles Treaty after First World War. This reality reinforces the alleged contractarian argument for uncompromising wars. Suppose you go to war with the aim of

---

[15] See Isabelle Duyvesteyn, 'Some Conclusions', in Jan Angstrom and Isabelle Duyvesteyn (eds.), *Understanding Victory and Defeat in Contemporary War* (New York: Routledge, 2007), 233.

[16] Johnson and Tierney show the decisive role of the media in creating an 'imagined victory' for the side that lost, as well as a sense of defeat for the side that won. They illustrate this claim by contrasting a case of military success perceived as a failure (Somalia) with a case of military failure perceived as success (Mayaguez). See Dominic Johnson and Dominic Tierney, 'In the Eye of the Beholder: Victory and Defeat in US Military Operations', in Jan Angstrom and Isabelle Duyvesteyn (eds.), *Understanding Victory and Defeat in Contemporary War* (New York: Routledge, 2007), 46–76.

[17] This insight poses a conundrum for contemporary democracies. On the one hand, they cannot win without projecting a clear image of victory. On the other hand, projecting such an image necessitates the infliction of destruction of a magnitude that democracies could not bear. This difficulty relates to a thesis developed by several writers about how strong states lose wars against much weaker adversaries. See Gil Merom, *How Democracies Lose Small Wars* (New York: Cambridge University Press, 2003), followed by Azar Gat, *War in Human Civilization* (Oxford: Oxford University Press, 2008), 627, and Ivan Arreguín-Toft, *How the Weak Win Wars: A Theory of Asymmetric Conflict* (New York: Cambridge University Press, 2005). On the more philosophical level of how *morality* might lose as a result of adherence to the (current) requirements of the morality of warfare, see Saul Smilansky, 'When Does Morality Win?' *Ratio* 23 (2010), 102–10.

averting a violation of the *ad bellum* contract and deterring the violator from further aggression. You must be willing not only to fight the war to its end, but also to strictly enforce the settlement that follows it; otherwise, your efforts and sacrifice will prove to be pointless. This is a bit like 'fighting' with your children. If you lack the requisite determination to enforce the rules at stake ('only one hour of TV a day'), it would be better to spare the fight and not have the rule in the first place.

The conclusion that seems to follow is that at times, the total destruction of enemy forces is a legitimate aim of war[18]—a conclusion that is hard to swallow. We believe that while contractarianism might suggest a right to go to a defensive war whose aim is total destruction of the enemy military forces, on reflection, it can explain how such a move can be resisted. The contractarian argument against uncompromising wars begins with a few considerations. First, while these wars are very detrimental to the side against which they are conducted, they are also pretty damaging to the side conducting them. They usually last a long time, cost a lot of money, and involve many casualties.[19] This supports the thought that, in most cases, uncompromising wars might not be as beneficial to those conducting them, as Clausewitz's view seemed to imply.

Second, a major consideration in favour of uncompromising wars is that a permission to fight such wars would deter states from further aggression. Our model presumes that while this might be true of some states, there is no reason to believe that it holds true in most of them. In more mundane contexts, like deterring students from violating school rules, no analogous 'uncompromising' measures are needed. More to the point, there is no reason to think that the death penalty for theft is the most efficient way to deter potential thieves. When partially decent states in minimally just symmetrical anarchy are concerned, suffice it that the measures impose some non-negligible price on the violators and that these measures be accompanied by explicit determination to repeat them and to make them harsher if necessary. Public opinion in decent states is flexible and sensitive to the costs individuals have to bear as a result of mistaken political decisions. In turn, the political decisions in these states are sensitive to public opinion. Hence, the model assumes that generally, to deter groups from future aggression, there is no need to 'annihilate' their military power *a la* Clausewitz, but only to undo their military achievements, to impose a non-negligible cost in casualties and in money, and to send a clear message regarding the defenders' determination to do the same—and more—in the future if necessary.

---

[18] See Rodin, 'Two Emerging Issues of *Jus Post Bellum*', 55, and Moellendorf, 'Jus ex Bello', section II.
[19] At surface value, some of these costs can be avoided by using nuclear weapons, which would achieve the required annihilation in a few minutes. But of course, the price that would be paid by the use of such weapons is enormous—both to the side using these weapons and to the international community in general. We address the legitimacy of using weapons of mass destruction in Chapter 7.

It follows that in minimally just symmetrical anarchy, uncompromising wars are not the optimal way to enforce the Charter's prohibition on the use of force. True, in some cases they definitely are (maybe the strikes against Nazi Germany and Japan in the Second World War would serve as an example). Should the war agreement simply leave it to each warring party to decide for itself whether to fight 'to the end' or to compromise?

We believe it should not. States should agree to subject themselves to a rule that categorically prohibits uncompromising wars between decent states. (Regarding wars against indecent states, see Chapter 7.) The rule is authoritative (in the Razian sense) because of the seductive nature of the Clausewitzian argument. The chronic lack of information about the enemy's capabilities, intentions, and future behaviour, coupled with the strong pressure from the populace, might motivate states to opt for uncompromising wars although this runs *against their own interests*. Moreover, leaders and public opinion tend to demonize the enemy once the war breaks out, so states tend to underestimate the prospects of peaceful solutions. This attitude might lead other countries to behave in a similarly harsh fashion, driven by the fear that if an actor behaves like a 'softy' in the jungle of international relations, his chances of being swallowed by the circling sharks increase. This, in turn, might lead to an escalation into a world in which uncompromising wars are not the exception but rather the norm. The significant risk of error and of escalation that follows should lead partial yet decent states to agree on an *ex bello* arrangement that outlaws uncompromising wars; an agreement to end wars without full certainty of being able to reap the political fruits of military achievements.

The argument for an *ex bello* regime that categorically prohibits uncompromising wars is structured similarly to the argument against preventive wars. In a decentralized society of states, each state must take care of its own interests and defend itself when it suspects that the contract prohibiting the use of force is about to be violated. Each state has a strong incentive to strike first and eliminate the threat. From a moral standpoint, such a preventive war would be desirable if it were less costly (in terms of innocent lives) than a defensive war to both sides. By the same logic, to prevent future aggression, countries may have a weighty reason to conduct uncompromising wars that could guarantee a reasonably long period of peace. The ultimate goal of uncompromising wars that such a regime would support is the prevention of the next aggressive war. To put the analogy differently, if a state would be justified in launching a defensive war at t2, why should it not go on to a preventive war at t1, thereby reducing the expected killing and harm? Similarly: if a state would be justified in launching a defensive war in, say, ten years, why should it not pursue the current war uncompromisingly, 'to its end', again with fewer casualties than would be incurred in that war at that future date?

Both questions invite the same answer; namely, that although pre-contractual morality allows some instances of preventive and of uncompromising war, partial and decent states have good reasons to tie their hands, so to speak, with a contract that rules out all such wars. By committing themselves to this rule, the sides to the contract reduce the chances of mistakes leading to suboptimal wars, and defend themselves against the grim prospect of being the object of preventive or uncompromising wars fought against *them*.

The contractarian argument for a rule that prohibits total destruction leaves open the exact point at time where the war ought to terminate. This point is radically indeterminate, and this indetermination is costly to both sides. One way to minimize it is to allow formal surrender, or another conventional act by which one side 'coerces' the other to terminate the war. These acts were quite common in the history of war, as we shall see in the next section.

4.4 Historical Agreements on Ending Wars

Some historical cases show that states enter agreements requiring the parties to cease fire short of a decisive military victory and definitely short of reaping the political benefits of war. According to legal historian James Whitman, eighteenth-century Europe provides an excellent illustration. This was 'an age of exceptional military restraint',[20] based on the mutual commitment of all parties to a set of rules about how to end wars and how to define victory. The battle of Malplaquet between the French and the Spanish is one of Whitman's central examples. In that battle, the French lost 11,000 men while the Spanish lost 21,000. The campaign against France had effectively failed. Nevertheless, even the French themselves perceived the battle as a Spanish victory. How was that possible?

> The answer belongs to the standard pre-modern law of victory, and it is an answer that suggests powerfully that there were rules in eighteenth-century warfare and that those rules mattered. Under the pre-modern law of victory, the French counted as the losers at Malplaquet because they were the ones who retreated... Malplaquet was a complete victory for the Allies because they managed to gain control of the field of battle, despite the fact that they did so only at the cost of devastating and hugely disproportionate losses.[21]

In this particular battle, it may seem odd to say that the 'rule of retreat' led to or expressed restraint. After all, Malplaquet was one of the bloodiest battles of the century. Nevertheless, the alternative—the continuation of fighting until the

---

[20] Whitman, *The Verdict of Battle*, 172.   [21] Ibid.

complete destruction of the enemy—would have been even worse. All the more so in less bloody battles, in which the retreat of one party from the battlefield was sufficient to terminate a war and determine the victor.

One might wonder what significance this artificial definition of victory may have had, given that the Allies suffered almost twice as many casualties as the French. To this, Whitman answers by showing 'that people in the eighteenth century took the question of who counted as the loser very seriously'. Being a loser in the technical sense of retreating from the field of battle had significant implications in propaganda, politics, and diplomacy. As a result, the losing party would be discouraged from launching war again in the near future, while the victor would experience the opposite: under the impression that its superiority was affirmed, the victor would be much more confident in reutilizing its military power in case it felt that doing so would advance its interests.

Defining victory in terms of driving the enemy from the battlefield makes warfare look like a game. This analogy did not escape Whitman's attention:

> When we consider the long history of the practice of pitched battle, it can be hard to resist the impression that a battle was a kind of game. A pitched battle, unlike a raid, pits two armed teams against each other. This has the air of a lethal team sport rather than a hunt. The winner in pitched battles like Malplaquet and Chotusitz was the side that forced its opponent off the field, no matter how high the human cost. This surely has the sound of something like a sumo match or a football game: the goal was to capture and hold the field, and your victory counted as victory even if your lineup was cruelly decimated...It is also noteworthy that a pitched battle takes place on a special field, just as so many organized sporting events do. The very word pitch is still used for sports fields, just as it was once used for battles...Pitched battle also includes one of the most striking features of organized sport: a kind of game clock, traditionally (if fictitiously) supposed to run from dawn to dusk.[22]

The idea of regarding pitched battles as a kind of lethal game is extremely helpful in illuminating their conventional nature.[23] It also connects to the boxing analogy mentioned in Chapter 2. Boxing rules determine not only what boxers are permitted to do to each other during the match (in an apparent violation of their opponent's natural right not to be harmed), but also when these permissions lose their validity (i.e. when the game is over). Like with the rules of war, boxing rules apply to three phases: they define when and how the match begins; when and how it ends; and how it should be conducted.

---

[22] Ibid.
[23] Whitman adds that 'it is a striking fact that humans sometimes literally do play games to resolve conflicts that could be resolved through war' (181).

Although the retreat rule played a crucial role in shaping eighteenth-century wars, it was not an invention of that time. As Whitman notes, 'in one way or another it was something of a Western universal for centuries'.[24] All the more so regarding the general phenomenon of rule-governed battles. In Azar Gat's view, face-to-face confrontation in pre-historical societies was very risky; hence the preference for the raid and the ambush as warfare tactics. When direct confrontation did take place, it was constrained by rules that significantly limited the violence, sometimes to the point of making victory merely ceremonial. Gat describes such confrontations in the hunter-gatherer era as follows:

> Conflict between clans or tribes could lead to face-to-face confrontations, or battles, the place and time of which were normally agreed upon in advance... The two opposing dispersed lines stood at a spear-throwing distance, about 50 feet, hurling spears at one another while dodging the enemy's spears. In some cases, such battles were intended in advance to put an end to a conflict and were thus truly 'ceremonial,' with the spear throwing restrained and mixed with ceremonial dances. Once blood was spilt, or even before, the grievances were seen as settled, and the battle was terminated.[25]

This mutual acceptance of rules regulating the conduct of battles and their endings had nothing to do with some kind of aversion to shedding blood. When some group estimated that it could crush its enemy—usually through a nightly raid, but at times also in the open—it did so mercilessly. Rather, self-restraint in battle had to do with mutual benefit; that is, with the high risk involved in direct confrontation and with the rational assessment that it would be better to avoid fatal injuries even at the cost of not completely defeating the enemy. As Whitman succinctly put it, 'it was in the interest of all the parties to accept conventions that defined victory in ways that stopped short of annihilation'.[26]

The ceremonial and game-like aspects of wars are less common nowadays than they were in the eighteenth century or in older wars, although they still exist.[27]

---

[24] Ibid., 183.

[25] Gat, *War in Human Civilization*, 117. See also 185: 'Formal battles were largely demonstrative, often producing more noise than blood.'

[26] Whitman, *The Verdict of Battle*, 197. See also his illuminating remark there, that Pufendorf regarded the law of war as a branch of contract law.

[27] A central purpose of Whitman's *Verdict of Battle* is to explain why conventions like the retreat rule gradually lost their appeal during the course of the nineteenth century and vanished altogether in the twentieth and twenty-first centuries. Part of the explanation he offers involves the fact that in the eighteenth century, as throughout most of human history, war was not thought of primarily as a horror, but as a legitimate procedure for resolving legal disputes. Because of this understanding of war, it was possible for the warring parties to restrain warlike activities according to well-defined rules, including rules that determined the war's end. For various reasons, this understanding was abandoned during the nineteenth century and replaced by a different one, according to which wars are fought to settle issues that have 'world-historical stakes'. When the stakes are high—a war against evil—little room is left for restrictions that might allow evil to continue to flourish.

A good illustration is found in the rules of surrender, which is commonly signalled by waving a white flag or raising open hands above one's head. When such symbolic acts are committed, the other side is expected to cease its fire immediately and be willing to negotiate a truce. The point we wish to emphasize is that these ceremonial acts of surrender impose inescapable limitations upon the side receiving them, even if it has a strong interest to continue the war so as to prevent future wars, or to reap the political fruits of its military success. *Ex ante*, the rules concerning surrender correspond well with the interests of all sides. Decent parties would prefer an arrangement that enabled losing parties to enforce a ceasefire on their rivals, even if that meant that if *they* were the triumphant side, they too would have to abide by these rules.

4.5 Deterrence and Prevention: The Just Aims of Defensive Wars

We are in a position to formulate a contractarian theory of the aims of just war and a contractarian approach to *jus ex bello*. The main idea underlying this theory is that just as the parties agree to avoid preventive wars, they also agree not to continue fighting in order to secure the prevention of all future aggressions. Once the enemy is thrown out of the territory it managed to occupy, or once it has given up its attempt to occupy the territory and is deterred from doing so in the near future, the war should usually come to an end. The contract does not demand the cessation of fire immediately after the blocking of an enemy attack, or immediately after removal of the just cause for war. As long as the enemy has not surrendered or has not made clear that it is willing to stop its aggression, fighting may continue. But the point is that if the enemy *has* taken such measures, the just side ought to quit the war and shift to diplomacy even if the military power of the enemy is still significant and even if there is no guarantee of long-term peace.

Our proposal, then, takes a middle ground between two extremes. One position says that fire must cease once the *casus belli* has gone. The other says that a state may keep fighting in order to make sure that the probability of future aggression is nearly zero. From the contractarian perspective, while states are not allowed to continue fighting in order to prevent all future wars with the aggressor, they are not required to be content with merely a temporal removal of the threat. They may continue fighting until the enemy explicitly expresses its willingness to cease its aggression.

As noted, while immature threats usually do not constitute a just cause for *going* to war, they might well be a just cause for *continuing* it. The central reason for this *ad bellum/ex bello* asymmetry involves the importance of efficient deterrence against potential violators of the contractual rule against initial use of force. Decent states should know that if they violate this rule, they will pay a price.

Not only will their enemy obtain a right to fight in order to remove the threat they pose, it will also obtain a right to hit them harder than is strictly required to defend its territorial integrity.

The permission to weaken the aggressor's army for the sake of deterrence implies another *ad bellum*/*ex bello* asymmetry. Going to war is justified only if it meets the necessity condition, which means that the victim of aggression is barred from using force before it makes sure that measures short of war, such as diplomacy and political pressure, cannot remove the threat. However, once the war breaks out, this duty is null and void. Furthermore, while fighting, the defender is exempted from the requirement to make sure every day anew that the fighting is still necessary in order to achieve the legitimate aims of the war.[28] The wide permission to weaken one's enemy once the war breaks out legitimates the right for the state that has been attacked to recruit all its military might to defeat its enemy, to undo the results of its aggression, and to deter it from using further force at least in the near future. It is under no duty to make sure that it is using only minimal force when it attacks the military forces of the enemy.

4.6 *Ad Bellum* vs *Ex Bello* Proportionality

The previous section pointed to several asymmetries between *ad bellum* and *ex bello* involving the *ad bellum* condition of last resort. In the present section, we address other *ad bellum*/*ex bello* asymmetries that are generated by two other *ad bellum* conditions; viz., a reasonable chance of success and proportionality. We start by briefly discussing probability of success and then turn to a more detailed discussion of proportionality. A war should not be fought unless its chances of success are reasonable. How should success be determined? We have shown the ambiguity and indeterminacy that surround attempts to determine who the real winner of a war is.[29] An attempt to predict more remote consequences of war is notoriously difficult as well. The contractarian theory of the just aims of war simplifies the moral issue. We suggest that the permission to use force in an attempt to deter aggressors and to prevent future wars implies a permission to disregard the question of which results count in assessing success. According to the contractarian view, the parties should judge military success in terms of blocking the enemy's aggression and deterring the enemy from using force

---

[28] For such a requirement, see Cécile Fabre 'War Exit', *Ethics* 125/3 (2015), 631–52.
[29] In Mandel's view, 'it is quite difficult...for most recent wars to generate widespread consensus that they ended unambiguously in either victory or defeat' ('Reassessing Victory in Warfare', 16). For a survey of the main dilemmas in this area, see Jan Angstrom's introduction to Angstrom and Duyvesteyn (eds.), *Understanding Victory and Defeat in Contemporary War* (New York: Routledge, 2007).

again in the near future (or for a significant period). If achieving these aims is improbable, the defender should hold fire.[30]

Let us turn to the *ad bellum* prohibition on disproportionate wars. Tom Hurka defines the proportionality condition as follows: 'Despite their differences, the various proportionality conditions—*ad bellum* and *in bello*, simple and comparative, objective and subjective—all say a war or act in war is wrong if the relevant harm it will cause is out of proportion to its relevant good.'[31] As shown in Chapter 1, many philosophers believe this understanding of proportionality to be incompatible with the blanket permission to go to war in defence of a country's borders. If, as Adil Haque puts it, '*Jus ad bellum* proportionality constrains each party's pursuit of military victory by comparing the value of their war aims with the cost of their pursuit', protecting territorial integrity by killing and maiming is disproportionate.[32]

Contractarianism thus offers a different understanding of the *ad bellum* proportionality condition, which we entitle 'contractual proportionality'. As we suggest, the premise underlying the agreement about proportionality is that the aim of the contract is to maximize defence of the pre-contractual rights of individuals without compromising the chances of the just side (the side whose contractual right not to be attacked has been violated) to win a contractually just (viz., defensive) war. To achieve this end—to enforce this mutually beneficial and fair contract—the parties agree to allow defensive wars even if most of them are pre-contractually disproportionate. Contractual proportionality is based on a further agreement. The parties agree on a different definition of the positive and negative effects that go into the proportionality calculation. The positive effect is the undoing of the violation of the contractual prohibition on using force and the weakening of the military forces of the aggressors, whereas the negative effect is the harm inflicted on civilians.

Now, put succinctly, contractual proportionality asserts that since wars are so costly in terms of harm to civilians, states should agree not to address small aggressions by means of war if the predictable costs of defensive wars to civilians are excessive. States should agree to find other means of dealing with such violations. Precisely because of the moral and prudential value of peace, there are crimes against peace that ought to be left unaddressed by force.

To see why partial but decent states should undertake such a proportionality condition, one should attend to their epistemic limitations once again. Given the perceived threat to their vital interests and against the background of uncertainty and vagueness, states (just like individuals) are prone to err in assessing the

---

[30] For possible exceptions, see Daniel Statman, 'On The Success Condition for Legitimate Self-Defense', *Ethics* 118/4 (2008), 659–86.
[31] Thomas Hurka, 'Proportionality and the Morality of War', *Philosophy and Public Affairs* 33 (2005), 38.
[32] Haque, *Law and Morality at War*, 161.

relevant facts and in interpreting their moral significance. They tend to overestimate ethical transgressions committed by their rivals, while underrating ethical failings of their own. On account of these well-known psychological biases, the parties agree that violations of the *ad bellum* contract that justify resort to war must be clearly visible to all relevant sides. This is why imminent threats to territorial integrity are the only *casus belli*. For the same epistemic reason, if the violations of the contract are minor, or if it is unclear whether the breaches will cause significant harm, the contract asserts that war is 'disproportionate'. The reasons for this are that (1) the smaller the breach, the greater the risk that a state misconceives it; and (2) the greater the predictable costs on the other side, the more reason there is to accommodate the violation and let it pass unaddressed by force. The contracting parties have an interest in reducing disproportionate defensive wars so defined. So they agree to undertake limitations on their *own* right to resort to war when the violation of their sovereignty is not that significant and when the harm required to block it is substantial, in order to allow themselves to engage in actions that the other side might consider a violation of the contract.

The contractarian redefinition of the positive and negative effects that feed into the proportionality calculation marks an important difference between contractarianism and revisionism. According to the revisionists, a war that cannot be justified on the basis of the regular, individualistic moral consideration necessarily fails the proportionality test. Hence, if a morally corrupt regime launches a defensive war, its war is probably disproportionate, because maintaining the political regime in question is simply not a good effect of the war, and as such cannot weigh in its favour. By contrast, according to contractarianism, when testing for the positive effects of war, all we need to look at is how likely it is that the defender's right against aggression will be protected, and how likely it is that the military forces of the unjust side will be weakened. Save extreme cases, a threat to territorial integrity of states is illegitimate regardless of the moral level of the invaded country. Consequently, removing the 'aggressive threat' comprises a positive effect—which is then to be weighed against the expected harm to civilians.

Now consider the role of proportionality in constraining the continuation of war. Suppose that at t1, State A estimates that going to war against State B would cost X casualties (including its own civilians and combatants, as well as the enemy civilians) and estimates that these expected casualties and harms are (contractually) proportionate to the goals of war.[33] At t1, the expected harm that needs to be caused in order to undo the right violation the aggressor causes is not very significant, so this violation should not go unaddressed (by violence). However,

---

[33] Whether or not these are the evils that the proportionality calculation must consider is controversial, but this issue extends beyond our scope of discussion here. For a helpful examination of the sorts of goods and evils that fall under proportionality in war, see Hurka, 'Proportionality in the Morality of War', sections 2–3.

the war does not go according to plan. At t2, X has been brought about, but the goals of war are not yet achieved.

One pre-contractual proportionality condition (which McMahan calls the 'Quota View') implies that, at t2, the war loses its moral justification and must come to an end. Since fighting a disproportionate war is impermissible and since, *by hypothesis*, the continuation of this war would make the entire war disproportionate—would exceed its 'quota'—it must stop. While McMahan rejects this view, Fabre seems right in regarding it as an implication of the individualist morality of war. True, there are various reasons for drawing a distinction between the pre-contractual proportionality of initiating a war and the pre-contractual proportionality of continuing to fight it. A country that ceases fighting once it realizes that its war was costlier than expected and therefore disproportionate would be in a worse position compared to the status quo ante. Typically, withdrawing from the battlefield after war has already been set in motion is almost inevitably interpreted as a sign of weakness, not as a heroic manifestation of a commitment to a pre-war proportionality judgment. Yet, these and other considerations are not sufficiently weighty. Fabre seems right in claiming that if killing in war must satisfy the standard conditions of legitimate self-defence, then insofar as the killing of any number beyond X violates the condition of (pre-contractual) proportionality at a certain at time, it is impermissible at all times.[34]

Yet, the quota view is 'implausibly restrictive'.[35] The players on the international level have a strong reason to seek exemption from it so as to enable deterrence and effective self-defence. For the same reason, they have good reason to exempt themselves from engaging in an ongoing proportionality calculation during the war. They realize how unstable and uncertain their assessments were prior to the war with regard to the expected course of war, the expected harm (to both sides), and so on.

Contractarianism would thus distinguish between proportionality as a condition for the very launching of war and proportionality as a condition for continuing to fight a war whose initiation satisfied the *ad bellum* conditions. The contractual basis for this distinction is clear. An agreement that requires the just side to withdraw from the battlefield, thereby letting the unjust side reap not only the direct fruits of its aggression (say some disputed territory that it unjustly seized), but also the strategic fruits of its apparent military victory, would lose its credibility. This potential outcome would undermine the aim of the parties to the contract to deter states from illegal use of force. Hence, on the one hand, it is in the interest of all states to agree to a list of constraints on the resort to war, including— as shown earlier—that of (contractual) proportionality. They agree to a proportionality condition that commands leaving minor violations of the prohibition on

---

[34] For Fabre's criticism of McMahan, see her 'War Exit', fn 9.
[35] As noted by Jeff McMahan, 'Proportionality and Time', *Ethics* 125/3 (2015), 702.

the use of force unaddressed (by force). On the other hand, once war breaks out, the contract allows the parties to calculate the proportionality of continuing the war periodically, rather than every day anew. More importantly, they agree that the proportionality test at any point of time during war will be purely prospective, exempting each other from the need to take past losses into consideration.

4.7 Conclusion

From time to time, states resort to war in order to defend themselves against real or imagined adversaries. To make their fighting effective, they often consider destroying their foes altogether, thus guaranteeing for themselves the political and economic benefits of their military success. The logic of this Clausewitzian argument is hard to resist, but it is incompatible with the accepted *jus ad bellum* regime.

Although uncompromising wars aiming at the 'annihilation' of the enemy military forces are usually bad for the states that fight them, in some cases they comprise the best option for the parties in the face of some perceived unjust threat. Absent the war agreement, uncompromising war might be morally justified. The problem is that states are prone to treat their situation as warranting uncompromising war even when it does not. To save themselves from such costly errors, decent states under conditions of minimally just anarchy give up the option of completely destroying their (decent) enemies. They agree to separate military victory from the achievement of long-term political benefits—mainly the prevention of any future aggression from their adversary. Such agreement to end war short of eliminating the risk of aggression is *ex ante* mutually beneficial to all parties, hence binding, even if *ex post* it runs against the legitimate interests of one of them.

The contractual distinction between the *ad bellum* and the *ex bello* phases is based on the impracticality of a constant re-evaluation of the prudence and the morality of a given war. The decision to commence war is not a decision to use limited force for one day or so, which would then be followed—or maybe not—by a similar decision to use force for another day or so, and so on. A full-scale war involves such an extensive mobilization of forces, such a moving around of military equipment, and such reorganizing of military and civilian infrastructure, that doing so just for one or two days would be unrealistic. To guarantee an effective right to defence from aggression, states have an interest in agreeing that a decision to go to war would completely change the moral landscape in the ways to which we have just alluded.[36]

This concludes our discussion of the conditions for starting and ending wars. We now turn to see how contractarianism helps to substantiate the rules of *jus in bello*.

---

[36] It is not surprising that a decision to go to war makes such a normative difference given the way decisions function in practical reasoning in general. See Joseph Raz, 'Reasons for Action, Decisions and Norms', *Mind* 84 (1975), 481–99.

5
Contractarianism and the Moral Equality of Combatants

5.1 Introduction

As explained in Chapter 1, the traditional *in bello* regime is comprised of three basic rules: (a) combatants may be attacked intentionally; (b) noncombatants may not be attacked intentionally; and (c) noncombatants may be attacked unintentionally if the attack is proportionate and an effort is made to minimize civilian casualties. These rules apply equally to Just and Unjust Combatants: Unjust Combatants may attack Just Combatants and be attacked by them, and both sides are allowed to inflict collateral damage on noncombatants of the other side.

Revisionists have strongly criticized the moral equality between Just and Unjust Combatants and between Just and Unjust Noncombatants. In their view, combatants fighting for the unjust side cannot claim a right to kill their adversaries in the battlefield, and definitely not to kill civilians as a side-effect of military operations. The purpose of the present chapter is to offer a contractarian defence of *Moral Equality*: the view that states that Just and Unjust Combatants have an equal moral right to kill and maim each other in war.[1] In accordance with the principles laid down in Chapter 2, we seek to show that under conditions of minimally just symmetrical anarchy, where the prohibition on aggression is enforced by self-help, the legal equality of combatants satisfies the conditions of *Mutual Benefit*, *Fairness*, and *Actuality*. The moral equality of soldiers is true by virtue of an agreement between states. By accepting it, combatants attain a right to take advantage of traditional *in bello* rules, thereby violating no duty against Just Combatants by killing them, even if these killings are pre-contractually impermissible.

Thus, the morality of killing in war is far more permissive than the individualist morality of self-defence. Notably, however, this morality is also more restrictive. If combatants violate these rules by targeting civilians and civilian objects, for

---

[1] It is not insignificant that historically, the emergence of the idea of moral equality involved a contractarian way of thinking. As Cheyney Ryan explains, the key figure in this emergence was Hugo Grotius, 'who contrasted the "law of nature" (classical conceptions of just war, and their Christian appropriation) with the "law of nations" which he saw as grounded *in the mutual consent of states*, as expressed in customary practices and international instruments like treatises' (see his 'Democratic Duty and the Moral Dilemmas of Soldiers', *Ethics* 122 (2011), 16, italics added).

example, they violate the right to life and the property rights of these civilians even if these civilians would be legitimate targets under pre-contractual morality. In this chapter, we defend the moral equality of combatants.[2] In the next chapter we defend the moral equality of civilians.

5.2 *Moral Equality* and the Importance of Obedient Armies

The guiding idea of contractarianism in general is that moral rights can be traded by accepting social roles. In the context of war, combatants lose their pre-contractual right not to be killed when they fight with a just cause but gain a right to kill enemy combatants when the *enemy* fights with a just cause. Combatants lose and gain these rights by subjecting themselves to the legal regime that confers on soldiers a legal right to fight a war without first making sure that the war they are fighting is morally justified. By accepting this legal regime, soldiers allow each other to undertake the duty of obedience; that is, to become instruments of their state.

As we noted in the introduction, other writers have proposed versions of this understanding as well. Walzer argues that soldiers are morally equal because 'military conduct is governed by rules [that] rest on mutuality and consent'.[3] Tom Hurka suggested that 'by voluntarily entering military service, soldiers on both sides freely took on the status of soldiers and thereby freely accepted that they may permissibly be killed in the course of war'.[4] Following these thinkers, we assume that as a social role, soldiery is shaped by treaty-based positive and customary international law, as well as by informal rules and widely shared attitudes. By enlisting in the military, combatants commit themselves to these rules—a commitment that is morally effective if the agreement that constitutes soldiery is almost universally accepted, and also mutually beneficial and fair.

The first task in defending *Moral Equality* is, then, to show that in minimally just symmetrical anarchy, the legal equality of combatants meets the terms of *Mutually Benefit* and *Fairness*. This section is devoted to this end.

Let us start by pointing again to the distinguishing feature of symmetrical anarchy. Usually, decent parties have a weighty prudential reason to bargain

---

[2] The versions of the contractarian approach to the *in bello* regime that we explore here were offered in Benbaji, 'A Defense of the Traditional War-Convention', 464–95, Yitzhak Benbaji, 'The War Convention and the Moral Division of Labor' *Philosophical Quarterly* 59 (2009), 593–618, and Benbaji, 'The Moral Power of Soldiers to Undertake the Duty of Obedience'.

[3] In fact, Walzer offers two models: 'The moral reality of war can be summed up in this way: when combatants fight freely, choosing one another as enemies and designing their own battles, their war is not a crime; when they fight without freedom, their war is not their crime. In both cases, military conduct is governed by rules; but in the first the rules rest on mutuality and consent, in the second on a shared servitude' (*Just and Unjust Wars*, 37).

[4] See Thomas Hurka, 'Liability and Just Cause', 210. See Christopher Kutz, 'Fearful Symmetry', in David Rodin and Henry Shue (eds.), *Just and Unjust Warriors*, 69.

rather than to fight in resolving conflicts between them. Hence, one might assume that the best way to advance the interests of all states would be to agree on mutual disarmament, thereby blocking the option of war altogether. However, in the absence of a universally recognized authority to enforce such an agreement and ensure that all parties do indeed disarm, consensual universal disarmament is impossible. The second-best alternative is a self-help based regime, under which states are allowed to use force against immediate threats to their sovereignty and territorial integrity.

The rule that equalizes Just and Unjust Combatants meets *Mutual Benefit* if a further plausible empirical assumption—hereinafter, *Obedience*-is maintained. In order for a regime that prohibits use of force and enforces this prohibition by self-help to work efficiently, soldiers in national armies must have a legal right to undertake the duty of obedience. That is, soldiers should have a legal right to participate in wars without first making sure that these wars are just. The empirical assumption that underlies the conviction that this legal order meets *Mutual Benefit* is this: it is in the interest of individuals who live in a minimally just symmetrical anarchy to be protected by states that control *obedient* armies. Any asymmetrical restrictions, like a right to fight just wars only, or a right to kill only Unjust Combatants, would undermine the main objective of the contract, which is to enable states to efficiently address ongoing aggression and to deter potential aggressors from unlawful use of force. Asymmetrical rules would compromise the obedience of combatants and thus the ability of states to act in self-defence.

Practically, everyone is expected to benefit from the national security that is gained due to soldiers' obedience—including the soldiers themselves. Like other members of the polity, combatants benefit from an *in bello* agreement that allows for their obedience because it makes national defence more efficient. Soldiers must pay a higher price to maintain it, but they do so for a relatively short period, until others replace them. Hence, obeying orders is not only permissible for soldiers, but as Joshua Greene noted in passing, it is a matter of virtue: 'Just as personal loyalty makes one a more attractive friend or lover, a disposition to respect authority can make one a more attractive foot soldier within a larger cooperative enterprise...Good foot soldiers have the virtue of loyalty and humility. They know their place and dare not abandon it.'[5]

Accordingly, states institute a regime under which combatants possess a legal right to participate in wars against other decent states, whatever their cause. Under this regime, soldiers are responsible only for their compliance with the *in bello* rules, not for the war itself. Legal equality further grants *post bellum* immunity to

---

[5] Greene, *Moral Tribes*, 43.

enemy combatants in order to secure similar immunity for the combatants of one's own side.[6]

Legal equality meets *Mutual Benefit* for a further important empirically-based reason. Any mutually beneficial *in bello* regime should satisfy the condition of self-help enforceability: the rules to which states subject their soldiers are such that make it possible to agree during the course of the war whether or not they were violated. This epistemic requirement is justified by the fact that the *in bello* rules can be enforced by the warring parties only. Consider, by contrast, wars governed by rules that allow the killing of only Unjust Combatants or only Culpable Civilians. Under such rules, combatants should be guided by *jus ad bellum* considerations. In wars governed by these asymmetric rules, each side would regard itself as entitled to retaliate for what it takes to be the enemy violation of the *in bello* code: each side would take the other as violating the rule that allows killing only Unjust Combatants, say. Such retaliations would aggravate the apparent injustice in the eyes of the other, which would lead to more violence, and so on, in a dangerous spiral. War governed by unenforceable rules is prone to escalate and increase the risks to both parties, without any compensating advantage. Thus, a regime under which soldiers should be guided by *jus ad bellum* considerations would bring about a Pareto inferior outcome compared to the outcome in which the regime allows soldiers to follow the orders of their political leaders.

Is the rule that equalizes soldiers an aspect of fair arrangement? Does it meet *Fairness*? According to our definition of fair contract, the answer is positive only if the arrangement neither *maintains* the unfair inequalities in the actual state of affairs nor creates unfair inequalities of its own.

We suggest that the legal equality of soldiers is a fair rule mainly because the *jus ad bellum* contract is a fair arrangement. As we argued in Chapter 3, an arrangement that prohibits the use of force (Article 2(4) to the UN Charter) and permits partial states to enforce this prohibition by self-help (Article 51) is fair because states would have accepted the 2(4) prohibition even if they were totally impartial. Indeed, states should accept this arrangement despite the fact that on rare occasions, weak parties can promote distributive justice only by violating the 2(4) prohibition on the use of force. In the absence of obedient armies, states would be less effective in blocking violations of the contract. Hence, an arrangement under which states hold obedient armies promotes the enforcement of *a fair* contract. It follows that under conditions of minimal justice, the *in bello* regime does not solidify unfair inequalities. Additionally, the contract does not create unjust differences between states or between individuals. The legal equality that

---

[6] Haque, *Law and Morality at War*, chapter 1, argues that international law confers merely legal *immunity* rather than a legal *right* to participate in an unjust war. This strikes us an implausible reading of the law.

the *in bello* contract commands cannot be known in advance to discriminate against any party in a symmetrical conflict between decent states. After all, the legal equality of soldiers is general and universal, and as such it cannot be known *ex ante* to *generate* unfair inequalities. (Although it does discriminate against stateless nations that have no armies. We will address this complication in Chapter 6).

Legal equality meets the terms of *Actuality*. It is nearly consensual that armies have a legal right to obey the orders they get from their respective political authorities. Likewise, the legal right of individual soldiers to obey their political leaders without verifying that the war they are fighting is just is accepted and practised everywhere.

5.3 Defending Obedience

How might one refute the claim that the *in bello* agreement that commands legal equality is mutually beneficial? The obvious option is to question the importance of obedience of soldiers in enforcing the prohibition on aggression. This is McMahan's approach. He argues that an asymmetrical regime that prohibits the participation of combatants in unjust wars would better protect the rights of states to sovereignty and to territorial integrity. He asks us to imagine how different things would be had Nazi soldiers avoided treating themselves as 'functionaries who have been given a job to do'.[7] In effect, McMahan argues, the rule of legal equality that empowers combatants to undertake the duty of obedience fails the *Mutual Benefit* test, as it runs against the interests of combatants:

> Potential combatants would have more reason to accept a principle that would require them to attempt to determine whether their cause would be just, and to fight only if they could reasonably believe that it would be. If they were to accept that principle, there would be fewer unjust wars and fewer deaths among potential combatants. Each potential combatant would be less likely to be used as an instrument of injustice, and less likely to die in the service of unjust ends.[8]

McMahan believes that the ignorance attributed to combatants is exaggerated. The information that enables them to judge whether a war is just or not is often accessible and sufficient. There was no need for a scrupulous investigation on anyone's part to determine that the Nazi invasion of Poland was unjust.

---

[7] McMahan, *Killing in War*, 101, quoting Stanley Milgram.
[8] Jeff McMahan, 'On the Moral Equality of Combatants', *Journal of Political Philosophy* 14/4 (2006), 384.

We disagree. Most cases are unlike the Nazi invasion of Poland; and given the deception and propaganda of the Nazi regime, even this war could not have been that easily identified as unjust at its very beginning by the young Germans sent to fight in it. Furthermore, while it may be true that combatants of liberal democracies can reach well-informed judgements about the morality of military campaigns, these judgements are formed under a regime that allows states to require obedience. States do not need to use intensive indoctrination in order to secure the obedience of their army. By contrast, if soldiers were not under a (legal) duty of obedience, states would adopt various forms of manipulation to convince them that they ought to fight. What follows is that in regimes that rejected the legal equality of soldiers, fewer combatants would probably refuse to participate in unjust wars than in regimes governed by this principle. Massive exposure to disinformation would lead soldiers to believe that their war was actually just.[9]

Now, we concede that it would be harder for states to fight unjust wars under an asymmetrical regime that requires soldiers to make sure that their war is just. We argue that under such a regime, it would be harder for states to fight just wars as well. Consider a pilot who is ordered to initiate a permissible preemptive war by launching a surprise attack against the army of an aggressive state. Suppose that the pilot suspects that the war she has been ordered to initiate does not have a just cause. Or suppose that she thinks that the war does not satisfy the last resort requirement. Under an asymmetrical regime, she would be entitled to refuse to participate in the attack unless she was convinced that her suspicion was baseless. But convincing her would require time and effort, and would make self-defence less effective and more costly. There is, therefore, a reason to believe that under an asymmetrical regime, it would be harder for just states to fight just wars and easier for unjust states (who are less likely to respect international law and more likely to engage in propagandist misinformation campaigns) to fight unjust wars.

In response, revisionists could claim that 'soldiers are disposed to trust the authority of their government... They are, moreover, less likely in general to judge that a just war is unjust than to judge that an unjust war is unjust.'[10] Hence, while an asymmetrical regime would have little effect on the willingness of combatants to participate in wars that turn out to be *just*, it might have a negative influence on their willingness to fight *unjust* ones. We doubt these speculations. Under a

---

[9] McMahan observes that 'the contemporary military organization that has the most conspicuous record both of instances of conscientious refusal to serve in certain campaigns... and of toleration of this sort of conscientious action is Israel's IDF'. He contends that this example supports the epistemic asymmetry between just and unjust wars: 'No one doubts that everyone in the IDF would fight with the utmost cohesion... in a just war of national defense' (*Killing in War*, 99). Clearly, however, in Israel, tolerance of conscientious refusal comes with a huge investment in convincing youngsters that Israel's wars are just, that being part of the IDF is honourable, and that avoiding military service is unfair and shameful.

[10] Jeff McMahan, 'Duty, Obedience, Desert, and Proportionality in War: A Response', *Ethics* 122 (2011), 145.

regime in which the authority of states over soldiers is compromised, soldiers who are required to join a just war might undertake a severe risk to their lives. They have an obvious incentive to judge the war they are asked to fight in as unjust and thus to spare themselves the risk involved in participating in it.

Despite the importance of the considerations offered above, the main objection to the asymmetric regime that McMahan envisions appeals to deterrence-related considerations: obedient armies allow decent states to deter potential aggressors. We speculate that the obedience of soldiers is important to achieving this goal. Suppose that armies were not under a legal obligation to obey the political leadership and to wage war if ordered to do so. Now suppose that the politicians want to deter the enemy by bluffing. They want to convince the enemy that their combatants will go to preventive or disproportionate wars that the contract forbids. Such a threat would be far more difficult to impose under a regime in which soldiers were forbidden from fighting preventive or disproportionate (and thus, unjust) wars. As a result, the power of deterrence would be significantly weakened. Since, under conditions of symmetrical anarchy, states have a strong interest in maintaining the ability to deter potential aggressors, they will accept a regime that facilitates obedient armies.

In sum, we offer three points in response to McMahan's objections to *Obedience*. First, under an asymmetrical regime, unjust states would invest more in propaganda to convince their combatants that their war is just. Hence, such a regime would reduce the number of unjust wars only marginally. Second, if states had to convince their combatants of the justice of their wars, fighting *just* wars would become costlier, thus making efficient self-defence more difficult. Third, and most importantly, under an asymmetrical regime, just states might lose the option of threatening (bluff-threats) to launch morally dubious wars.

Of course, the justification for obedient armies is connected to the idea of the moral division of labour. In states that are basically decent, only the government or the president has the authority to start a war by issuing orders to the army to do so. The army has no legal right to make such a decision, regardless of what it thinks about the threats posed by neighbouring countries. Also, as indicated in the previous chapter, just as the initiation of wars is the job of politicians and not of the army, so is their ending. It is not the role of the military—certainly not the role of individual combatants—to decide on the termination or continuation of hostilities. Morally, *as soldiers*, they need not ask themselves whether the aims of war have been realized or not. Certainly, they ought not to make their continued participation in war dependent on their answer to this question. In the moral division of labour, the responsibility to end wars is allocated to politicians.

Contrast this view with a morality of killing and maiming in war that adopts *Individualism* and *Continuity*. This approach rejects *Moral Division*, as well as the idea of a contract that trades the basic rights that accompany it. Since individualists reject the moral division of labour between politicians and combatants, they impose on each individual combatant the responsibility to continually consult the principles of *jus ad bellum* to make sure that they justify each act of warfare she is ordered to carry out. The same applies to *jus ex bello*. The decision of individual combatants to continue fighting should be subject to the same principles governing the initiation of war; namely, *jus ad bellum*. This burden imposed upon members of the military might seem attractive if one believed, as does James Pattison, that 'if a few more individuals begin to question the permissibility of their contribution, this would be a positive development'.[11] But Pattison's optimism seems unrealistic to us. An individual combatant who stops fighting because she believes that the war should come to its end is not on higher moral grounds than a prison guard who decides to let some convicted murderer go free because, in his estimation, she has suffered enough. Such an act is no more socially desirable than that of a policeman who disobeys an order to arrest a suspect because he believes her to be innocent. Social life would be unstable if individualist advice were taken seriously.

There is another reason for scepticism about the positive outcome of combatants deciding on principles of individual morality whether to begin or to end war. True, in some cases such personal initiative might lead some combatants to refuse to participate in wars that, objectively speaking, happen to be unjust. But it might just as well lead to the opposite outcome—to combatants *initiating* acts of warfare, or to continuing a war even when their government has decided to cease fire.[12] If ordinary soldiers are morally required not to fight an unjust war even if this means disobeying orders delivered by the political leadership that represents them, then the converse must also be true. That is, they ought to wage a just war, even if they are not ordered to do so.[13] But, as Dan Zupan points out, the main function of the *jus ad bellum* condition of proper authority is to prohibit private military action. Revisionists cannot fully embrace this *ad bellum* restriction.[14] One reason to oppose private wars (by no means the only one) is that they are likely to be unjust. Similarly, given the 'deeply ingrained can-do attitude that often leads [professional combatants] to believe that "one more push" can turn around a situation',[15] leaving decisions about the continuation of war in the hands of each individual combatant is dangerous. This is partly because such individual decision-making capacity runs a higher risk of continuing unjust wars than would be the case if such decisions were allocated to politicians. Thus, there is no reason to

---

[11] James Pattison, "When Is It Right to Fight? Just War Theory and the Individual-Centric Approach', *Ethical Theory and Moral Practice* 16 (2013), 53.
[12] For this problem of 'symmetrical disobedience', see McMahan, *Killing in War*, 92, and Ryan, 'Democratic Duty and the Moral Dilemmas of Soldiers', 36–9.
[13] Cheyney Ryan, 'Moral Equality, Victimhood and the Sovereignty Symmetry Problem', in David Rodin and Henry Shue (eds.), *Just and Unjust Warriors*, 151–2.
[14] Dan Zupan, 'A Presumption of the Moral Equality of Combatants: a Citizen-Soldier's Perspective', in Rodin and Shue (eds.), *Just and Unjust Warriors*, 215–17.
[15] David Rodin, 'Ending War', *Ethics and International Affairs* 25/3 (2011), 361.

think that the spread of individual morality would reduce rather than increase the number and the duration of unjust wars.

Is it true that in most cases, the division of labour that empowers soldiers to be obedient has a Raz-like legitimate authority over its subjects? The answer must be negative, in light of the fact that most wars between decent states have been unjust. Soldiers have no reason to believe that they fight for a just cause and/or that their war is proportionate and necessary, *just because* their leaders ordered them to join this war. Hence, we have not committed ourselves to the Razian authority of the rule that allows soldiers to follow leaders' order to go to war. We merely argued that the arrangement that exempts soldiers from verifying that their war is just is mutually beneficial, especially because maintaining obedient armies helps states to deter potential aggressors.

As we have emphasized, the contractarian argument works even if an arrangement is mutually beneficial but has no Razian legitimate authority over its addressees. Social rules are morally effective if they are freely accepted in the relevant society. Particularly, the arrangement that equalizes the legal standing of soldiers so as to empower them to undertake the duty of obedience may be morally effective even if states have no legitimate authority over soldiers with respect to the justice of their wars. *Moral Equality* follows from the fact that the legal equality of combatants is mutually beneficial—it helps states to deter their potential enemies from aggressing against them—and as such it is presumably true that states and the individuals they represent freely accept it.[16]

5.4 Do Combatants Accept Legal Equality?

According to contractarianism, when people within generally decent societies undertake social roles—as police officers, bankers, executioners, or combatants—morality confers on them the rights that are associated with these roles. By becoming combatants, soldiers accept the principle of legal equality and waive their right not to be killed by their adversaries. The obedience of soldiers and the legal equality that follows from it are a publicly recognized, integral part of the profession of armies, just as locking inmates in their prison cells is a publicly recognized, integral part of the profession of prison-guards and just as caring for the interests of clients is an integral part of the banking profession.

McMahan concedes that in joining the army, combatants undertake the duty to protect their country and, by implication, the risk of being unjustly but legally

---

[16] Does the statistical fact that most wars in history were unjust favour refusals? Not necessarily. The fact that most marriages fail should not have any impact on your assessment of the prospects of your own marriage. Note also that refusal might be justified, according to our argument. An arrangement that exempts soldiers from verifying that their war is just does not obligate them to fight this war in case they know that it is unjust.

attacked. However, he argues, taking risks does not amount to waiving rights: 'A person who voluntarily walks through a dangerous neighbourhood late at night assumes or accepts a risk of being mugged; but he does not consent to be mugged in the sense of waiving his right not to be mugged, or giving people permission to mug him.'[17] Moreover, the undertaking of the *in bello* code can be understood with no reference to rights at all. Imagine, suggests McMahan, a Polish man, enlisted in 1939, arguing as follows:

> There is a convention that combatants should attack only other combatants who are identified as such by their uniform. It is crucial to uphold this convention because it limits the killing that occurs on both sides in war. I will therefore wear the uniform to signal that I am someone the convention identifies as a legitimate target. In doing this I am not consenting to be attacked or giving the Nazis permission to attack me; rather, I am attempting to draw their fire toward myself and away from others.[18]

Thus, there is no reason to think that by joining the army, combatants free their adversaries from the duty not to unjustly attack them. If the latter are not freed from this duty, then insofar as they fight for an unjust cause, they have no moral right to kill their enemies—in contrast to *Moral Equality*.

In response, we suggest that the nature of social roles transcends the self-understanding of any individual within or beyond the institution. It is like the implications of marriage, which depend not on the subjective perception of the partners, but on the social definition of the institution and the bundle of rights, duties, and liberties that it entails. Thus, to determine what holders of a role should do, it is not necessary to explore the inner minds of combatants who undertake the duties that define their role. One should look instead at the legal sources that define this role—positive and customary international law—and the way they are commonly understood. Once an individual becomes a combatant, she thereby consents to the terms of the role so defined. Joining military forces is, in other words, an act 'such as participation, compliance, or acceptance of benefit that *constitutes* tacit consent to the rules of an adversary institution'.[19] The institutional norms that define this role emerge from the social structure within which this role is created and maintained. Once again in Hardimon's words, 'what one signs on for in signing on for a contractual social role is a package of duties, fixed by the institution of which the role is a part'.[20]

One might still argue that the soldiers' consent to the institution of soldiery is morally effective only if they are properly informed about the implications of such consent. Arguably, since most of them know little international law and do not

---

[17] McMahan, *Killing in War*, 52. [18] Ibid., 55.
[19] Applbaum, *Ethics for Adversaries*, 118. [20] Hardimon, 'Role Obligations', 354.

understand the distinctive moral contours of their role, their decision to join the army cannot ground the waiver of important rights. But the marriage example again casts doubt on this objection. The lesson that this example conveys is that even if at the time of marriage both individuals hold the erroneous belief that their marriage creates an indissoluble relationship, they nonetheless have a right to exit the marriage according to the conditions specified by the law. The formal acceptance of a social role is an authentic instance of consent to the norms that define it, even if detailed acquaintance with the legal boundaries of the relevant institution is lacking.

Of course, the very description of a wedding ceremony implies that the participants are aware of some aspects of the process they undergo and of the relations that are created by engaging in it. If the husband is 'systematically deceived, so that during the marriage ceremony he believed that he was undergoing a mysterious ritual to become a Freemason',[21] he cannot be said to have consented to be married; therefore, he cannot be described as having waived any moral right he possessed. Under normal circumstances, however, individuals know that they are getting married although they are not always aware of all the normative implications that follow. Returning to soldiery, if a person enlists into the military 'in the belief that he is joining the Freemasons', then indeed, he has not consented to become a soldier and cannot be said to have undertaken the duties associated with the profession of soldiery. Nor can it be said that he has waived his natural right not to be killed unjustly by enemy combatants. But this is hardly ever the case.

This analysis is inspired by an externalist intuition from the philosophy of language. In order to refer to water by pronouncing the word 'water', the speaker need not know what water *is*; namely, she need not know that water is H_2O. Still, she does have to be aware of some typical properties of water, for instance, that it is a transparent, drinkable liquid.

A closely related concern is that any consent-based argument for the moral standing of the *in bello* rules applies only to combatants who freely took on their status as combatants, not to conscripts. But this worry is also misguided. First, the fact that conscripts are required by law to join the army does not imply that they do so unwillingly. They might fully identify with the law and find it justified. Second, even individuals whose enlistment into the army was a result of legal coercion can plausibly be said to have consented to the role of soldiery. As Margaret Gilbert argues, although in one sense the reluctant conscript signed up against his will, in another sense, which she calls the 'decision-for' sense, his enlistment was not involuntary; the enlistee may have decided to join the army in order to avoid punishment.[22] Similarly, as opposed to the victim of a pickpocket, a

---

[21] McMahan, 'Duty, Obedience, Desert, and Proportionality in War: A Response', 149.
[22] Margaret Gilbert, 'Agreements, Coercion, and Obligation', *Ethics* 103/4 (1993), 685.

person who hands over his wallet to a mugger offering a choice between his money or his life *coercively consented* to the transfer. Following Gilbert, we assume that, despite duress, the reluctant conscript accepts the rules that define the role he occupies because he decided to join the army.

To drive this point home, think of somebody who is coerced by his circumstances into entering a boxing ring and fighting against some adversary. He does so because he's desperate for the money to save his own life or the life of his child. Nonetheless, once he enters into the ring, he thereby undertakes to conduct himself according to the rules of the game. He can hit his adversary only within the range defined by these rules and cannot complain when his adversary (who is unrelated to his plight) hits him in return. The fact that his participation in the boxing match is a result of coercion does not invalidate his consent to be 'unjustly' attacked by his opponent.

This line of argument relaxes one of the conditions that a regime should meet in order to be morally effective. We argued that the fact that such a regime is mutually beneficial makes it presumably true that those who habitually obey it freely accepted the rules to which the regime subjects them. Free acceptance, we assumed, entails waiver of rights. But, we now suggest, when it comes to *explicit* acceptance, the requirement for moral effectiveness is weaker. If one undertakes the package of norms by explicitly accepting a role, and the rules defining the role are mutually beneficial, fair, and habitually followed, then one thereby undertakes the duties, gains, and rights associated with the role, even if the acceptance of the role was coerced.

One reason that this modification sounds plausible is related to the fact that practices meeting the conditions of *Mutual Benefit* and *Fairness* create legitimate expectations among the relevant parties. When some individual wears a uniform and identifies herself as a combatant, she thereby creates expectations regarding her moral and legal status. In particular, she creates the expectation that she is subject to the regime that regulates warfare, within which all those wearing a uniform are legitimate targets for lethal attack. It will not help her to say that she had no intention of creating such expectations, just as it will not help a boxer to say something like that when entering the ring under well-defined circumstances.

This argument does not deny the existence of some leeway for societies in constructing the role of a combatant. In particular, while states are entitled to subject their combatants to the duty of obedience, they are free to allow their combatants to refuse to participate in a war that they find unjust. What the contractarian view denies states is the right to treat *enemy* combatants as criminals. It asserts that the moral right of combatants to undertake the duty of obedience is a direct implication of the right granted to states to have obedient armies. States cannot benefit from the right to maintain obedient armies while treating the obedient combatants of their enemies as criminals—thus, in effect, denying those benefits to their enemies.

Another objection to the contractarian defence of *Moral Equality* runs as follows. According to the current *in bello* regime, if an aggressive army invades Nicaragua and is opposed by combatants of the Nicaraguan Army, the invaders violate no right in killing these combatants. If, instead, they unjustly invade a country that has no army, like Costa Rica, where they are countered by a coordinated defence by people who take up arms as individuals, their acts of killing are impermissible. The problem is that this mere organizational difference between the defensive forces of Nicaragua and Costa Rica seems insufficient to explain the huge moral difference between the two cases.

We bite the bullet. As shown in detail in Chapter 2, it is such 'mere organizational' aspects that have a direct impact on the distribution of moral rights and duties in all fields of social activity—universities, banks, prisons, and so on. In all the above fields, societies have much to benefit from the establishment of the roles in question, but such roles come at a price too; namely, vulnerabilities that would not apply to the same individuals engaged in the same act absent such a role. Thus, while states have a lot to gain by the right agreed upon to hold obedient armies, the price they pay is in the soldiers' moral and legal vulnerability to attack (by the enemy army), which would not have applied to the same individuals qua regular civilians. When soldiers fight against soldiers, they are released from the moral requirement to make sure that the latter are liable to defensive attack as (*inter alia*) initiators of an unjust threat. When they fight against civilians, they are not released from this (pre-contractual) requirement.

Nothing in what we have said contradicts the fundamental idea that individuals have a natural—and asymmetrical—right to defend themselves and others from unjust aggression. The point is that they cannot exercise this right by choosing to fight as partisans if they are part of a society that takes advantage of the right to institute an obedient army. If the society to which they belong decides to fight through recruiting combatants, then combatants and civilians become subject to a package of legal norms that changes their respective moral standing. Qua combatants, individuals do not fight as individuals against other individuals, but rather fight under a set of rules that define their role. In turn, civilians should respect a division of labour in which they are not part of the military campaign.[23] When a political group does not have an army, or when its army is disbanded (e.g. following its defeat), then civilians regain their right to fight for their security and political autonomy.

---

[23] Which is why Matt, in the example we offered in Chapter 1, does not enjoy the same rights as his soldier brother John. See also Fletcher and Ohlin, *Defending Humanity*, 180, explaining why a Polish farmer in 1939 would be barred from opening fire on German troops.

5.5 Contractarianism and Treachery

Contractarianism explains why even Just Combatants are legitimate targets, and it explains why *all* combatants are legitimate targets (regardless of their actual contribution to the war effort or to their moral responsibility). In addition, it sheds light on some particular rules concerning warfare that might otherwise seem mysterious. Here is one example:

WHITE FLAG: Bob is a Just Combatant fighting with his platoon against enemy forces. They cannot identify the source of the fire shot at them. Bob has an idea. He'll come out of his shelter with a white flag and walk slowly towards the estimated location of the enemy. Then, when enemy soldiers come out to take him prisoner, Bob's comrades will kill them.

Would it be acceptable for Bob to do this? Almost everybody would answer this question in the negative and moreover would be disgusted by the mere idea. This is the paradigm of *treachery*. But given that Bob is fighting—as you recall—for the just side, this sense of revulsion is not that easy to understand. To see this point more clearly, consider:

DECEIVING HOSTAGE: Tamar is taken hostage with ten other innocent civilians by two terrorists who threaten to execute a hostage every hour if their demands are not accepted. She decides to act in order to save her own life and the lives of the others. She pretends to have fainted and when one of the terrorists comes over to help her, she snatches his pistol and shoots him and his friend.

Here, we assume that most readers would support Tamar and think that she acted bravely and admirably. Yet, as per *Individualism*, the difference between these two cases is incomprehensible. After all, both Tamar and Bob faced an unjust threat to their lives. So if Tamar is allowed to use deceit in order to overcome her kidnappers, why is Bob not allowed to do the same to overcome enemy soldiers who are unjustly maiming and killing his comrades (and collaterally killing civilians as well)? If, as *Individualism* has it, fundamental human rights are independent of affiliation or role, why should Bob be barred from doing to his attackers what he *would* have been entitled to do to them had he not been in uniform?

We believe that contractarianism provides a better way to think about such cases. The reason that Bob is not allowed to play this trick is that by accepting the role of soldier, he accepted the rules associated with it, including the rules against treachery and perfidy, which means that he waived his pre-contractual right to use such methods in war and, moreover, that he undertook a commitment not to do so.

To repeat a point made in Chapter 2, it would be misleading to describe Bob's dilemma as one between his natural right to deceive his pursuer in order to save his life[24] and his obligation to obey the law.[25] If that were the case then surely his right to defend his life from his (assumedly unjust) pursuer would take precedence. The law in this area defines not only legal duties but moral ones as well. To violate the law against treachery is not just wrong in some technical sense; it is an act of treason.

5.6 Responsibility for Killing Just Combatants

According to contractarianism, by accepting the rules that define their profession, combatants lose (or successfully waive) their moral claim against being unjustly attacked by enemy combatants. In response, individualists argue that even if the rules of war are mutually beneficial and fair, and even if combatants freely accept them, Unjust Combatants have no moral right to kill combatants who are justly defending themselves, their families, and their country. Since the killing of innocents is a fundamental violation of the moral standing of individuals, consent cannot render it morally legitimate. Neither tacit nor express agreement can change the fundamental moral standing of human beings.[26]

Yet contractarianism seeks no such change. It treats the killings carried out by Unjust Combatants just as it treats the killings carried out by 'unjust' executioners who carry out mistaken verdicts issued by the courts. When an innocent inmate is unjustly executed, his right to life is violated; but it is the state, rather than the executioner, to whom this violation should be ascribed. Similarly, the rights of Just Combatants who are killed and maimed in aggressive wars are violated by the aggressive state rather than by the individual combatants who carry out the aggression. The regime that allows for combatants' obedience is thus respectful, because when combatants waive their claim against enemy combatants not to be unjustly attacked by them, they consent to a regime that locates the responsibility for such an attack on states.

This conception of responsibility clarifies the difference between the regime sustained by the rules of war and the practice of duelling.[27] Under the regime that

---

[24] Of course, not everybody would agree that people have such a right. See Immanuel Kant, 'On a Supposed Right to Lie From Altruistic Motives' in Lewis W. Beck (ed.), *Critique of Practical Reason and Other Writings in Moral Philosophy* (Chicago, IL: University of Chicago Press, 1949), 346–50.

[25] The relevant law is article 37(1) of the 1977 Additional Protocol I to the Geneva Convention: 'Acts inviting the confidence of an adversary to lead him to believe that he is entitled to, or is obliged to accord, protection under the rules of international law applicable in armed conflict, with the intent to betray that confidence, shall constitute perfidy.'

[26] To repeat (see chapter 2, fn. 44), Tadros rejects the view that combatants have a right to harm each other because they consented to being liable to be harmed in the course of a war. In his view, 'consent is neither necessary nor sufficient to create a conflict of permissibility' (*The Ends of Harm*, 114).

[27] See Chapter 2 above, Section 2.5.

allows for duels, it is the responsibility of the duelling individuals to make sure that the duel in which they participate is used to enforce vitally just claims. As a judge of his own case, a person who challenges another person to a duel is the sole entity responsible for his actions. In contrast, the social meaning of the role occupied by executioners and combatants is such that their agency is conceived as the medium through which the state acts. The division of moral labour that creates these roles transfers to the state the responsibility for any wrongdoing committed by role-holders in reasonably executing their roles.[28]

The significance of combatants' right to disregard the first-order reasons pertaining to the justness of their cause should not be overstated. First, as citizens, Unjust Combatants do carry some responsibility for the wrongs committed by their states—no more so than other citizens, but no less either. Second, as a matter of psychological fact, agents feel a special connection to their own actions and their results, which is distinct from the way they relate to the actions of others. This logic applies with special force to killing. Normal individuals engaged in the killing of human beings cannot avoid suffering the moral emotion famously described by Williams as 'agent-regret', an emotion that gives them further reason to check the justness of the war they are called upon to fight.[29]

Third, the fact that role-holders in general and combatants in particular have a right to act within the rules defined by their profession does not mean that they are under a duty to do so even if they are convinced that overall, such an act would be morally unjustified. Thus, we do not rule out the moral option of refusal on the part of role-holders—be they combatants, executioners, or policemen. Moreover, it is probably desirable that soldiers who are sure that their war is unjust do not take advantage of the right that they possess to participate in it. Note, though, that under conditions of minimally just symmetrical anarchy, which are already arranged in accordance with a division of labour, role-holders rarely have access to all the first-order reasons that pertain to the issues under their jurisdiction and rarely have the time and the ability to gather this missing information.

5.7 Conclusion

The purpose of this chapter was to deal with a troubling challenge to the traditional *in bello* code; namely, that it grants combatants of both the just and

---

[28] Note that vicarious liability of employers for their employees' actions is based on a different arrangement. When vicarious liability is imposed, it is the role-holder who violated the rules that define her role; and yet, her employer is responsible for this violation. Under the arrangement described here, soldiers follow the rules that define their role. Since their agency is the medium through which the state acts, it is the state that is responsible for their actions.

[29] Bernard Williams, 'Moral Luck' in Daniel Statman (ed.), *Moral Luck* (New York: SUNY Press, 1993), 42.

the unjust sides an equal right to kill each other. It seems morally outrageous that Unjust Combatants have a right to kill Just Combatants.

In response, we argued that within a contractarian view of morality, this arrangement makes perfect sense. Decent and self-interested states under conditions of minimally just symmetrical anarchy would prefer such an arrangement to an asymmetrical code that permits killing only Unjust Combatants, especially because of the importance of obedient armies and the dangers involved in asymmetrical *in bello* rules. Without armies whose combatants are entitled to follow the order to participate in war, decent states would not be able to efficiently enforce and protect the contractual prohibition on use of force. As suggested in Chapter 2, an arrangement by which decent states are governed is morally effective if it satisfies three conditions: *Mutual Benefit*, *Fairness*, and *Actuality*. We sought to show that under the circumstances of minimally just anarchy, the legal equality of soldiers satisfies all of these conditions. If so, then by accepting this arrangement soldiers generate *Moral Equality*; viz., a symmetrical distribution of rights and duties between Just and Unjust Combatants. Within the well-defined conditions of warfare, combatants of one side waive their claim-right not to be attacked by combatants of the other. This means that Unjust Combatants are not merely excused for killing Just Combatants; they are under no duty towards the latter not to attack them.

As we clarified, contractarianism does not imply that soldiers who know that their war is unjust must participate in it. To the contrary, *if* (unlikely) soldiers happen to know that their war is unjust, then they probably ought to refuse to take part in it. The contention that soldiers do not wrong enemy combatants by fighting an unjust war is consistent with the claim that they ought not to participate in the war for other moral reasons. Furthermore, contractarianism does not rely on the assumption that if decent states decide to go to war, the war is probably just. (In fact, there is no reason to assume that typically, decent states tend to go to just wars and to avoid unjust wars.) Rather, soldiers are legally exempted from making sure that the war that they are commanded to fight is just. The legislator assumes that in a world governed by a war agreement that equalizes their legal standing, there will be fewer wars than there would have been in a world governed by pre-contractual morality.

In this chapter, we showed how the *in bello* agreement affects the rights of combatants. In the next chapter, we discuss how it affects the rights of civilians.

6
Contractarianism and the *Moral Equality* of Civilians

6.1 Introduction

As we have developed it so far, *the jus in bello* code is comprised of two main parts. One regards the killing (and harming more generally) of combatants, while the other is concerned with inflicting harm on civilians. In the previous chapter we dealt with the liberty-right to target enemy combatants regardless of whether they belong to the just or the unjust side, of the actual threat they pose, and of their personal responsibility for the war in which they participate. The present chapter is devoted to the morality of killing and harming civilians in war.

We shall call the basic rule crystallizing the 'general protection from the effects of hostilities' *Civilian Immunity*. International law holds that 'in order to ensure respect for and protection of the civilian population and civilian objects, the Parties to the conflict shall at all times distinguish between the civilian population and combatants and between civilian objects and military objectives and accordingly shall direct their operations only against military objectives'.[1] Most fundamentally, *Civilian Immunity* requires that '[t]he civilian population as such, as well as individual civilians, shall not be the object of attack'.[2] Clearly, intentionally attacking individuals whom the attacker knows or believes are civilians is strictly forbidden. *Civilian Immunity* further requires that attackers avoid mistakenly attacking civilians, if such a mistake can be avoided. Finally, collateral harm caused to civilians must be necessary: '[i]n the conduct of military operations, constant care shall be taken to spare the civilian population, civilians and civilian objects'.[3] If it is possible to spare them while achieving the legitimate military aim, sparing them is obligatory. For instance, '[w]hen a choice is possible between several military objectives for obtaining a similar military advantage, the objective to be selected shall be that the attack on which may be expected to cause the least

---

[1] The last quotes are from Protocol Additional to the Geneva Conventions of 12 August 1949, and relating to the Protection of Victims of International Armed Conflicts, 8 June 1977 (Protocol I) art 48. quoted in Haque, *Law and Morality at War*, 106.

[2] Protocol I art 51(2)., quoted ibid.

[3] Protocol Additional to the Geneva Conventions of 12 August 1949, and relating to the Protection of Victims of International Armed Conflicts, 8 June 1977 (Protocol I) art 57(1), quoted in Haque, *Law and Morality at War*, p. 154.

danger to civilian lives and to civilian objects'.[4] As Walzer notes, attackers must both *not try* to kill civilians and *try not* to kill civilians.[5]

Finally, *Civilian Immunity* imposes a proportionality constraint on the permission to harm civilians collaterally, as a necessary side-effect of targeting military objects. The proportionality principle prohibits attacks 'which may be expected to cause incidental loss of civilian life, injury to civilians, damage to civilian objects, or a combination thereof, which would be excessive in relation to the concrete and direct military advantage anticipated'.[6] As Adil Haque comments, the law regulates 'how armed forces may pursue a particular military advantage' by requiring that they do whatever possible to spare civilians. Through the proportionality principle, the law also regulates 'whether a particular military advantage may be pursued or must be abandoned' if civilians cannot be spared.[7]

Civilian Immunity is often explained in terms of pre-contractual morality. As Henry Shue puts it, the principle is thought to be

> a reaffirmation of the morally foundational 'no-harm' principle. One ought generally not to harm other persons. Noncombatant immunity says one ought, most emphatically, not to harm others who are themselves not harming anyone. This is as fundamental, and as straightforward, and as nearly non-controversial, as moral principles can get.[8]

The 'no-harm' principle itself is conceived as an aspect of non-consequentialist morality; rare cases aside, it is morally impermissible to harm innocent people intentionally, even if such an act could prevent greater evil.

Yet as we noted in Chapter 1, viewed from the deontological perspective that underlies the no-harm principle, the principles that prohibit the targeting of civilians and allow their collateral harm present serious difficulties. First, it is unclear why all civilians should be regarded as innocent in the relevant sense; namely, as immune to defensive attack. After all, many of them—particularly politicians, but also some officials, civil servants, and scientists—are as morally responsible for the war their countries wage as combatants are, or even more so.[9] Second, the Doctrine of Double Effect (DDE) that is supposed to substantiate the

---

[4] Protocol I art 57(3).
[5] Walzer, *Just and Unjust Wars*, 155–6.
[6] Protocol Additional to the Geneva Conventions of 12 August 1949, and relating to the Protection of Victims of International Armed Conflicts, 8 June 1977 (Protocol I) art 51(5).
[7] Haque, *Law and Morality at War*, 175.
[8] See Henry Shue, 'War', in Hugh LaFollette (ed.), *The Oxford Handbook of Practical Ethics* (Oxford: Oxford University Press, 2003), 742.
[9] See Cécile Fabre, 'Guns, Food, and Liability to Attack in War', *Ethics* 120/1 (2009), 36: 'Although noncombatants are often thought to encompass all civilians, the latter (as has often been noted) often participate in the war: as citizens, they sometimes vote for warmongering political leaders; as taxpayers, they provide the funds which finance the war; as journalists, they can help sway public opinion in favour of the war; as political leaders, they take the country into war. Last, but not least, as workers, they

permission to inflict proportionate collateral damage on civilians faces serious challenges that 'have probably reduced it to a minority position among moral philosophers'.[10] As we have seen in Chapter 1, the distinction between manipulative and eliminative killing has also fallen short of explaining the distinction between cases of direct attacks on civilians and cases of collaterally causing them harm.

In light of these difficulties, this chapter proposes a contractarian account of the prohibition against targeting civilians and of the permission to collaterally harm them. *Civilian Immunity* and *in bello* proportionality are explained as terms in a mutually beneficial and fair contract into which decent states under conditions of minimally just symmetrical anarchy should enter. The parties should find a way to enforce the prohibition on the use of force while minimizing casualties in doing so.

In Section 6.2 we show that under an arrangement by which states defend themselves and deter others from aggression against them by holding armies, a contract granting immunity to civilians from direct attack is mutually beneficial to all decent parties. In Section 6.3 we argue that this contract satisfies *Fairness*. Sections 6.4–6.5 offer a contractarian reading of the permission to inflict collateral damage on civilians and an interpretation of the *in bello* necessity and proportionality rules. Section 6.6 explores whether civilians freely accept their legal standing in wars and whether their acceptance of the legal system is morally effective. Section 6.7 addresses the rules of warfare in asymmetrical conflicts. In particular, we explore whether the asymmetrical nature of such conflicts also entails asymmetry in the rules to which the parties are subject.

6.2 *Mutual Benefit*

The prohibition against targeting civilians and the requirement to take all measures to spare them while targeting military objects is not grounded in an intrinsic moral difference between civilians and combatants.[11] Rather, this notion is grounded in a valid contract between decent states that distributes the right not to be lethally attacked in ways that might well differ from the natural, pre-contractual distribution of this right. This interpretation of *Civilian Immunity* is

---

provide the army with the material resources without which it could not fight, such as weapons, transports, construction units, but also food, shelter, protective clothing, and medical care.'

[10] McMahan, 'Intention, Permissibility, Terrorism, and War', 345. McMahan adds that when he was one of the editors of *Ethics*, between 2000–07, he reviewed a number of submissions in which DDE was 'relegated to a footnote and dismissed as an exploded view that no reasonable person could take seriously'.

[11] Sections 6.1–6.3 revisit and modify certain aspects of the argument in Benbaji, 'Justice in Asymmetric Wars: A Contractarian Analysis'.

plausible first and foremost because, under conditions of minimally just anarchy, a mutually beneficial war agreement would include such a term. *Civilian Immunity* enables states to minimize the harm caused by wars, as part of an agreement that secures their ability to wage defensive wars and deter potential aggressors. Put differently, decent states should accept the *Civilian Immunity* restrictions because, typically, these restrictions will not jeopardize their ability to achieve effective self-defence against unjust threats posed by an enemy that usually respects the war agreement.

Let us explain. As we interpreted the war agreement so far, (obedient) national armies comprise the means by which states defend themselves and deter other decent states from violating the contractual prohibition on the use of force. The consensual aim of armies in wars is the defeat of enemy military forces in a manner that makes further aggression much less likely in the near future. Therefore, in a world in which this arrangement is respected, deliberate attacks on civilians are usually insufficient for winning wars between decent states. In addressing aggression, an army must deal with the military forces of the enemy. Attacking other enemy targets, such as politicians or civil servants, will not usually promote this aim. Attacks on civilian objects are insufficient even if the responsibility that these civilians bear for the war is no less than that of enemy combatants who were currently carrying out the aggression. The same is true of attacks on buildings that house these individuals—government buildings or the offices of newspapers that support the (perceivably) unjust war. The odds that such attacks on civilian infrastructure would weaken the army are usually pretty low. They hardly ever save the need for engagement with the military units of the enemy. And, to repeat, without weakening the army, blocking aggression and deterring aggressors are unlikely.

Now, we do not deny that attacks on civilians and on civilian-related infrastructure might make a significant contribution to military victory, under other arrangements. We merely observe that if the war agreement as described up to this point is accepted, this is usually not the case.[12] We suggest, in other words, that the *in bello* contract applies to a world in which the war agreement is accepted. In this world, a nation at war must first of all deal with the military threat against it; hence, typically, attacking enemy civilians would not accomplish much. (Note that when we talk of the expected (lack of) benefit from attacking civilians or civilian-related infrastructure, we do not refer to facilities that are directly involved in the

---

[12] For a biblical illustration of how an attack on an enemy's civilian population can determine a battle, see the description of the war against Ai in Joshua 8:19–20: 'And the ambush arose quickly out of their place, and they ran as soon as he had stretched out his hand: and they entered into the city, and took it, and hasted and set the city on fire. And when the men of Ai looked behind them, they saw, and, behold, the smoke of the city ascended up to heaven, and they had no power to flee this way or that way: and the people that fled to the wilderness turned back upon the pursuers. And when Joshua and all Israel saw that the ambush had taken the city, and that the smoke of the city ascended, then they turned again, and slew the men of Ai.'

enemy military effort, such as factories for the manufacture of bullets and weapons or centres for the transmission of military communications. The military value of attacking targets like these is often significant; hence, it would be difficult to reach an agreement on refraining from doing so.)

Thus, the morally relevant loss from granting immunity to civilian targets in (symmetrical) wars between decent parties is usually slight, as long as both sides adhere to it. Yet compared to the baseline—viz., an outcome in which decent parties try to follow pre-contractual morality—the gain (to both sides) from this rule is substantial. Absent *Civilian Immunity*, wars would have been bloodier. Moreover, if the parties follow the rule, civilians on both sides can continue to lead 'normal' lives while the warring armies are maiming and killing each other on the battlefields, often far away from the civilian centres. Therefore, decent parties have a strong reason to exit the 'state of nature' (that is governed by pre-contractual morality) and agree on an *in bello* contract that includes *Civilian Immunity* as one of its basic conditions.[13]

Now, if attacking civilians and civilian-related infrastructure is less effective than attacking soldiers and makes little contribution to victory, it would be impermissible by pre-contractual standards as well. Why should any decent state agree to immunize civilians whose responsibility for the war is unquestionable, if killing them is in any case pointless, and therefore impermissible? The answer is that, sometimes, targeting these civilians *is* helpful. Why would decent yet partial parties undertake a rule that prohibited targeting civilians in *all* cases, even where terrorizing the enemy by killing officials or politicians might be helpful? Why not stick to the pre-contractual rule according to which such targeting depends in each case on the expected utility of doing so?

For two good reasons. First, the parties recognize their proneness to err in times of conflict, especially once the war starts. They realize that when confronted with the perceived successes of their enemy and with an increasing number of casualties, they will face pressure to 'do something' about this and might choose morally dubious methods even when such methods have a low chance of success. They realize that when faced with a serious threat to soldiers or to civilians, we tend to 'an inward focus dominated by short-term gains. We direct our attention toward satisfying our innate needs, which are driven by self-preservation.'[14]

Second, the parties recognize the danger of escalation. If one side carried out attacks on targets located in its enemy's towns, its enemy would feel a need to retaliate in kind—which would lead to counter-retaliation, and so on. To prevent

---

[13] For a very different empirical analysis, see Steven P. Lee, *Ethics and War: An Introduction* (Cambridge: Cambridge University Press, 2012), 202–7. Lee believes that the best way to minimize the harm caused by wars is to end them as quickly as possible, which requires the removal of the prohibition against attacking civilians; total war is the preferred moral option.

[14] Bazerman and Tebrunsel, *Blind Spots: Why We Fail to Do What's Right and What to Do About It*, 71.

such escalation, it is in the interest of all parties to restrict the category of 'military' (hence legitimate) targets to targets that are undeniably involved in the military effort. Factories that manufacture tanks are fair game, but nothing more distant than that. The parties should restrain themselves in their attacks even when the expected benefit seems significant. If only targets that are clearly related to the war effort are attacked, it is harder for the attacked party to justify retaliatory attacks on targets that are not clearly related to the war effort. For these two reasons, it is beneficial to agree to a blanket prohibition against direct attacks on civilians that limits the permission to attack non-military objects to those clearly related to the war effort. Therefore, if the arrangement is fair (we will discuss conditions of fairness shortly) and freely accepted by the parties, then even culpable civilians whose killing might promote a just victory attain the right against being targeted in war.

A further objection reads as follows. If the goal of the *in bello* agreement is to reduce the harm and death brought about in war (without precluding victory), why stop at civilians? The objector does not argue for an agreement that bans wars altogether, which was discussed briefly above.[15] She imagines an agreement according to which the victor is determined by a very limited violent confrontation. The well-known proposal made by Goliath in I Samuel 17 illustrates this thought experiment. The armies of the Israelites and the Philistines were on the verge of a massive confrontation when Goliath

> stood and cried unto the armies of Israel, and said unto them: Why are ye come out to set your battle in array? Am not I a Philistine, and ye servants to Saul? Choose you a man for you, and let him come down to me. If he be able to fight with me, and to kill me, then will we be your servants: but if I prevail against him, and kill him, then shall ye be our servants, and serve us.

Seemingly, the Goliath agreement—to coin a term—is far superior to the current *in bello* one.[16] It offers both sides an effective way of protecting their vital interests; that is, killing the adversary in a duel. It does so by reducing the harm to and killing of both combatants and noncombatants to almost zero. It seems, *ex ante*, that both sides have an equal chance of producing a talented warrior who could defeat the warrior put forward by the enemy.

However, the Goliath agreement is unfeasible because its implementation by a legal system would require a complete abolition of armies. This is because, if states trust that their potential enemies will respect it, armies would be rendered unnecessary. (They would institute an international force to deal with outlawed

---

[15] See Section 3.2 above.
[16] See Mavrodes, 'Conventions and the Morality of War', 125, who refers to the proposal in the *Iliad*, book 3 to settle the war over Troy by a single combat between Paris and Menelaus.

states and non-state organizations.) In other words, the Goliath agreement requires pacifism; a world in which international conflicts are not resolved by war, but by a very limited competition between two individuals, one from each side. (Actually, one wonders, why go for a Goliath-style duel and not for a chess competition, or, to expedite the decision, simply toss a coin?)

Like a pacifist regime that requires mutual disarmament, the Goliath regime would be hopelessly unstable.[17] Given the absence of a common power that could enforce it and make states confident that their potential (decent) enemies are militarily powerless, decent states have no reason to believe that their decent enemies will disarm themselves. Hence, given states' interest in defending their pre-contractual and contractual rights, no state would give up on its national army. And if states continued to hold armies, they would almost certainly use them in the face of perceived unjust threats. They should not believe that their enemies will let a duel decide their fate.

It follows that the relations between decent nations are doomed to be plagued by suspicion and mistrust for the simple reason that despite their decency, nations prioritize the defence of their rights over the defence of the rights of others. Therefore, the only stable arrangement is the one that treats these facts as given. The best feasible arrangement allows for armies and predicts that aggression is a permanent danger in international relations. The war agreement tries to reduce the horrors of war by a convention that licenses using all the necessary military power a state has at its disposal to achieve military victory, while immunizing as many innocents as possible.[18] In fact, in light of the instability of an agreement of mutual disarmament, most states in effect have a *duty* to maintain armies as part of their obligation to protect their citizens from unjust threats. States with no army might fail to uphold the most fundamental obligation they have toward their citizens—to provide national security.

The same logic applies to attempts to impose restrictions on the development of new kinds of weapons or of platforms like fully automated robots.[19] The extent of moral concern caused by these robots is controversial.[20] Even if they are morally

---

[17] See Mavrodes, ibid., 127, estimating that 'there is almost no chance that such a convention [like the Goliath one] would actually be followed'.

[18] We believe that this provides a satisfactory answer to the question that troubles Norman in *Ethics, Killing and War*, 164: 'why is it that some restrictions on warfare stand some chance of being accepted [like that against attacking civilians]...whereas others such as the "single combat" convention do not?' *Contra* Norman, there is no need to assume that if some restriction is undertaken successfully, it is because this restriction has 'some independent moral weight' (ibid., 165).

[19] See, for instance, Noel Sharkey, 'Saying "No!" to Lethal Autonomous Targeting', *Journal of Military Ethics* 9 (2010), 369–83 and Kenneth Anderson and Matthew C. Waxman, 'Law and Ethics for Autonomous Weapon Systems: Why a Ban Won't Work and How the Laws of War Can', American University Washington College of Law. Research Paper No. 2013-11. Available at http://papers.ssrn.com/sol3/papers.cfm?abstract_id=2250126.

[20] See Daniel Statman, 'Drones and Robots: On the Changing Practice of Warfare', in Seth Lazar and Helen Frowe (eds.), *The Oxford Handbook of Ethics and War* (Oxford University Press, online edition, 2015).

concerning, however, given the absence of a common power, and given the mistrust that permeates international relations—states in the process of developing such robots should not abandon this project on the understanding that other states that command similar technological capacities would do the same.

6.3 *Fairness*

Is an *in bello* contract that permits killing combatants but prohibits killing civilians fair in a case where the citizens are responsible for the unjust threat against which the just side is fighting? Does such a contract solidify or generate unjust inequalities?

Civilian Immunity is formally egalitarian, as it imposes the same constraints on all decent states under conditions of symmetrical anarchy. Presumably, formal equality implies that it cannot be known *ex ante* that following this rule will *create* unjust power relations between the strong and weak states *that hold national armies*. Yet, as we shall see in Section 6.5, it can be known in advance that a stateless decent people might become even weaker if they subject themselves to *Civilian Immunity*. A stateless nation can better defend itself from aggression by a war that is regulated by pre-contractual morality.

It might be argued, however, that while *Civilian Immunity* does not generate inequalities, it maintains oppression of weak states by powerful states just like the contract between slaves and masters (where slaves undertake the duty to comply with their masters' orders, while, in return, the masters undertake the duty to avoid violent coercion). Furthermore, one might claim that *Civilian Immunity* systematically prevents disadvantaged states from achieving justice in the distribution of resources, opportunities, etc. Both concerns emerge from a simple observation: our argument does not deny that on occasion (albeit rarely), fighting a war regulated by pre-contractual morality might change an unjust status quo.

We already addressed a similar concern with respect to the legal equality of soldiers (in Section 5.2). We argued that if the *ad bellum* contract does not solidify unjust inequalities, neither does the *in bello* rule that equalizes combatants, provided that legal equality is the most efficient means by which the *ad bellum* contract is enforced. One might think that a similar argument applies also to the rule that equalizes *non*combatants, namely, to the justification of *Civilian Immunity*. But this is not the case. To see why, let us remind ourselves of the structure of the above argument. One task of the *ad bellum* contract is to secure the ability of decent states to address illegal use of force by conferring a right of self-defence. We showed in Section 3.4 that under conditions of minimally just symmetrical anarchy, the *ad bellum* contract (which prohibits wars whose aim is the implementation of pre-contractual justice but permits defensive wars whose aim is the enforcement of this prohibition) is fair. It does not generate power inequalities, and typically does not prevent the worse-off party from permissibly achieving a fairer distribution of resources and opportunities. Hence, under such circumstances, the *in bello* contract, whose aim is the efficient enforcement of the *ad bellum* contract, is also fair: this arrangement is the most efficient feasible means by which states can enforce a fair *ad bellum* contract.

Civilian Immunity faces a distinct fairness concern, however. Suppose a weak state learns that in order to address an illegal use of force by a powerful aggressor, it must deliberately kill a few officials who are responsible for it. Subjecting itself to *Civilian Immunity*, the weak state refrains from such killing and, consequently, fails to address the violation of its rights. *Civilian Immunity* seems to solidify unfairness by preventing the weak side from defending its (contractual) right to self-defence.

In response, we propose that the *in bello* arrangement is fair in a different sense than the one delineated earlier. True, even if both sides observe them, the constraints imposed by *Civilian Immunity* might deny the warring sides various opportunities for military advantage that they could enjoy by fighting a war in which they were free to treat civilians according to the verdicts of pre-contractual morality. Still, the arrangement is fair if the chances of victory that the parties have in a war not committed to *civilian Immunity* do not systematically change because of features like military might or economic wealth. Conversely, an *in bello* contract would be unfair if it systematically changed the chances of victory in comparison to a war governed by pre-contractual morality. An *in bello* contract whose acceptance causes such a change might reduce the ability of the just party to enforce its contractual right against aggression.

Why presume that *ex ante*, *Civilian Immunity* does not systematically change the chances of victory? Under the *ad bellum* arrangement, states attain effective self-defence by significantly weakening the military forces of the aggressor. So, typically, a state's chances of victory depend on its military strength. Killing civilians has no impact on this matter. Thus, we may assume that if both sides respect *Civilian Immunity*, their chances of winning a limited war would typically be identical to their chances of winning an otherwise identical war whose *in bello* code treats civilians as pre-contractual morality does. Hence, typically, the chances of the just side to block aggression will not change on this basis. *Civilian Immunity* is fair in the sense that it benefits both sides and that it does not systematically compromise the chances of the *just side* to defend its contractual claim against aggression. *Civilian Immunity* might rarely and contingently solidify an unfair disadvantage (that was gained by a violation of the *ad bellum* code); but even such an occurrence does not undermine the fairness of the war agreement if, indeed, it is unpredictable through features like prosperity, size, etc.

This notion of fairness operates in another context. Consider the rule that requires boxers to wear padded gloves. Each is thereby able to inflict less harm on the other, but each also suffers less at the hands of the other. This rule is advantageous both to the stronger boxer and to the weaker one. The stronger boxer has nothing to lose but much to gain from this convention. He is better protected from injuries that might be caused by a gloveless adversary. The weaker boxer will, of course, also gain from this convention. Given his inferior position vis-à-vis his rival, he would be much more vulnerable if the latter took off his gloves (*Mutual Benefit*). Moreover, the gloves rule maintains the power inequality. There is no reason to think that boxing without gloves would improve the weaker boxer's odds or vice versa (*Fairness*). This analogy explains why, if only one side observes the *Civilian Immunity* constraints, the other side will be significantly advantaged, and why its chances of victory might change significantly. The analogy explains why the parties are tempted to violate *Civilian Immunity*, and why the war agreement should include a mechanism to guarantee that both sides follow the *in bello* rules. We present this mechanism in the next chapter.

Under this conception of *Fairness*, the *in bello* agreement that grants immunity to civilians and civilian society is unfair when applied to an asymmetric war of independence, where the weak side can target civil society but cannot effectively threaten military objects. As we shall see in Section 6.5, in those cases, the *in bello* regime does seem to favour the strong party and might well maintain unjust inequalities.

6.4 *Collateral Damage*

We have argued that under conditions of symmetrical anarchy, the war contract assumes that refraining from attacks on civilians usually does not hinder the chances of victory, and therefore forbids intentional attacks on them and on civilian infrastructure. The *in bello* contract further assumes that this argument cannot be extended to collateral damage to civilians. Combatants may bring about harm to noncombatants as a side-effect of attacks on military targets, provided that the harm is necessary and not disproportionate to the military goal and that efforts are made to minimize civilian casualties. We have called this principle *Collateral Damage*.

The war agreement assumes that in many cases, harming civilians as a side-effect of attacks on legitimate military targets is necessary to achieve crucial military objectives. It assumes that crucial targets are often located in proximity to residential areas, and that, consequently, it would be hard to fight effectively without causing harm to civilians. In particular, states fighting just wars may have no choice but to use air power to defend their just claims. But 'with aerial warfare civilians became extremely vulnerable and were inevitably collateral targets, potentially on a much larger scale than previously'.[21] A rule that immunizes

---

[21] See Judith G. Gardam, 'Proportionality and Force in International Law', *American Journal of International Law* 87 (1993), 399.

civilians from collateral damage would imply a limitation of the scope of national armies' right to defend the just claims of their states. Unlike attacks directed at civilians, attacks directed at military targets that unintentionally harm civilians are often instrumental in advancing military victory. Therefore, so we suggest, the parties to the war contract agree that such collateral damage would be allowed.

Note that the contractarian distinction between incidental and deliberate harms to civilians is very different from the pre-contractual distinction between manipulative and non-manipulative killing or between intentional and foreseeable killing. The parties assume that while deliberately attacking civilian targets is usually unhelpful in defeating one's enemy, attacking military targets is—even when it involves the collateral killing of civilians. Hence, like the restraining aspects of the *in bello* agreement, *Collateral Damage* comprises part of an agreement that partial yet decent states undertake in order to facilitate effective self-defence, on the one hand, while reducing the horrors of war, on the other.

This interpretation strongly suggests that civilians are under contractual duty to avoid using force. Pre-contractually, whoever is threatened by the aggressive campaign has a right to defend him- or herself. If a group of people is unjustly attacked, then all group members seem to have a right to oppose their attackers by force. (They might, of course, choose to entrust the task of doing so to some individual or sub-group.) If so, pre-contractual morality implies that if an army unjustly threatens them, civilians are allowed to set up ambushes or carry out other warlike activities against their enemy. Our interpretation suggests that civilians are under a contractual duty to distance themselves from the military forces in various ways.[22]

One of the aims of the *in bello* contract is to reduce the horrors of war by a strengthening of the separation between combatants and noncombatants; between the battlefield and civilian life. Since the active involvement of civilians in warlike activities undermines this separation, states have a strong interest in ruling it out. This would be problematic if refraining from civilian involvement undermined the possibility for effective self-defence. But if the war agreement is accepted, it can be safely assumed that it does not; the contribution of unorganized civilians to the war effort is usually marginal at best, in a society of states in which state self-defence is enforced by armies. Hence, a rule forbidding civilians from taking part

---

[22] As we noted above (chapter 1, fn. 48), whether or not civilians have a positive legal right to fight is disputed. Those who believe that they have a legal liberty-right to fight agree that if they do attack enemy soldiers and are caught, they may face criminal charges for murder or other offenses. Unlike soldiers, they are not immune from legal prosecution by a foreign state. Thus, Haque quotes with approval the International Committee of the Red Cross, stating that 'civilian direct participation in hostilities is neither prohibited by IHL nor criminalized under the statutes of any prior or current international criminal tribunal or court' (*Law and Morality at War*, p. 25). Compare however, Eyal Benvenisti and Doreen Lustig, 'Taming Democracy: Codifying the laws of War to Restore the European Order', 1856–74 (unpublished ms), who, regarding the codification of the laws of war in the nineteenth century, argue that 'instead of laws to protect civilians from combatants' fire, the laws that were drafted sought to protect combatants from civilians'.

in war—and, less controversially, denying them immunity from legal prosecution—is mutually beneficial. Civilians should be seen as having given up their pre-contractual right to use force against unjust attackers invading their country. The agreement that makes all soldiers legitimate targets in war excludes civilians from this game. By accepting it, they waive their right to take arms. This contractual constraint on civilians is the flip side of their contractual immunity from (direct) military attacks.[23]

6.5 *In Bello* Necessity and *In Bello* Proportionality

The *in bello* proportionality condition follows smoothly and naturally from the basic idea that drives the liberty-right to collaterally harm civilians as a side-effect of targeting military objects. The proportionality rule constrains this permission. It prohibits the achievement of military advantage by inflicting excessive collateral damage to civilians. We suggest a contractarian interpretation of *in bello* proportionality that relates it to the general aim of the *in bello* contract; namely, to reduce the horrors of war without preventing effective self-defence. Before presenting it, we discuss two preliminary questions that our contractarian interpretation of the *in bello* proportionality rule is intended to resolve.

The first fundamental question reads as follows. Under its standard understanding, an act is proportionate only if its (relevant) positive effects outweigh its (relevant) negative effects. Yet, no individual act that (alongside others) constitutes an unjust war has positive effects that can appropriately be weighed in the proportionality calculation. As McMahan puts it, 'how...could a Nazi combatant weigh the harms he would cause to enemy combatants (or to enemy noncombatants) against the end of victory by the Nazis without assigning any value to the victory?'[24] McMahan concludes that 'it is rather mysterious what traditional just war theorists have been assuming in their supposition that Unjust Combatants can satisfy the requirement of proportionality in the same way that Just Combatants can'.[25] After all, the military advantage that Unjust Combatants achieve by attacking military targets has only negative (instrumental) value and, therefore, cannot possibly be proportionate. Similarly, argues Hurka, 'no act by combatants on a side without a just cause can satisfy [*in bello*] proportionality: if their acts produce no expected goods, they can never be just'.[26] A symmetrical permission to bring about proportionate collateral harm to civilians seems conceptually impossible.

---

[23] For the idea that it is the responsibility of states to maintain a separation between their military facilities and residential areas, see W. Hays Parks, 'Air War and the Law of War', *Air Force Law Review* 32 (1990), 1–226.
[24] McMahan, 'The Ethics of Killing in War', 715.   [25] Ibid., 717.
[26] Hurka, 'Proportionality in the Morality of War', 44.

This criticism of the traditional view suggests an asymmetrical arrangement, according to which while Just Combatants may inflict collateral damage on enemy civilians, Unjust Combatants would be denied the right to inflict any harm on civilians. At face value, such an arrangement would achieve the optimal balance between the desire to facilitate effective protection against aggression and the desire to reduce harm to innocents.

Contractarianism explains why decent states would reject an asymmetrical proportionality rule. Succinctly put, it should not be a term in the war agreement that regulates their military conflicts. This is because such a rule cannot be enforced by self-help, and therefore, in an agreement where one of its terms is asymmetrical, proportionality would fail the mutual benefit test. Indeed, a rule can be part of the *in bello* regime only if the warring parties could agree during the war on whether the rule was violated or not. Combatants almost always believe that they are fighting on the just side and accordingly that the war fought by their enemy is unjust. Hence, under an asymmetric arrangement—where only the just side has a right to cause collateral damage—any aerial bombardment exercised by one side and bringing about the collateral death of civilians is likely to be perceived by the other as a violation of *in bello* proportionality; as the unjustified slaughter of innocent people. This perception would invite retaliation against civilians of the other side, and so on, in a cycle of lethal escalation. An asymmetrical *in bello* agreement would thus increase the violence of the war rather than diminish it. Therefore, states should prefer a symmetrical agreement that would grant all warring parties the right to cause collateral damage to civilians of the other party.[27]

The other question we should discuss before sketching a contractarian *in bello* proportionality rule regards its standard formulation: as it is understood in the Law of Armed Conflict (LOAC), proportionality prohibits targeting a military object if the attack is expected to cause civilian losses that are excessive in relation to the military value of the attack. The question is: what does it mean for a military attack to have 'military value'?

The most natural answer (an answer that we will shortly reject) is the following. Since the parties to the contract assume that state self-defence will be achieved by military victory, military value might be considered to be a function of causal contribution to victory. The greater the contribution to victory, the higher the military value of attacking the target; hence the broader the permission to kill and injure civilians as a side-effect of the attack. In deliberating whether to target a military object, the state should make sure that the attack is sufficiently important to a weakening of the enemy army, otherwise killing enemy soldiers is not

---

[27] For further discussion, see Yitzhak Benbaji, 'The War Convention and the Moral Division of Labour', *Philosophical Quarterly* 59 (2009), 597–8.

necessary, and therefore impermissible. After all, there seems to be no *intrinsic* value in destroying some army base or in killing some enemy combatants.

Note, however, that under the current regime, armies are allowed to destroy the military forces of the enemy without making sure that doing so is necessary to achieve the legitimate aim of the military campaign. As Gabriella Blum notes, the principle of military necessity permits 'only that degree and kind of force... that is required in order to achieve the legitimate purpose of the conflict, namely the complete or partial submission of the enemy at the earliest possible moment with the minimum expenditure of life and resources'.[28] Or, as the International Committee of the Red Cross (ICRC) recognizes in its Interpretive Guidance, '[a]part from the prohibition or restriction of certain means and methods of warfare, the specific provisions of IHL do not expressly regulate the kind and degree of force permissible against legitimate military targets'.[29]

What could ground such a sweeping permission? Why shouldn't combatants be required to establish that the attacks they carry out significantly promote victory? The answer offered by contractarianism is that imposing upon soldiers such a requirement would render national self-defence almost impossible: assessing the degree to which eliminating a military object promotes victory or secures deterrence from further aggression is typically time-consuming and unlikely to be successful.[30] Assessing the causal connection between particular attacks and the achievement of victory is usually nearly impossible. Wars often go on for quite a long time and are comprised of thousands of discrete attacks on combatants, vehicles, bases, and so on, with no realistic way of determining—in advance or even *post factum*—what the causal contribution of each of these attacks to victory might be (or might have been).

We suggest, therefore, that, through their states, combatants subject themselves to an agreement that exempts all warring sides from the deliberative duty to make sure that killing them promotes victory. Soldiers accept an arrangement under which they may be targeted in war *merely by having the status of soldiers* rather than under the regular conditions that govern the right to attack people in self-defence. This arrangement implies a conception of military value: an attack has military value if its object is part of the war machine. The destruction of this object has military value, even if it turns out that this object had no role whatsoever in the enemy's actual warfare: it could be an ineffective soldier, or a rocket that had no chance of being used.[31] The parties agree that since, in most cases, actions like

---

[28] UK Ministry of Defence, 2004.

[29] Gabriella Blum, 'The Dispensable Lives of Soldiers', *Journal of Legal Analysis* 2/1 (2010), 78.

[30] For an elaboration on this argument, see Daniel Statman, 'Can Wars Be Fought Justly? The Necessity Condition Put to the Test', *Journal of Moral Philosophy* 8 (2011), 435–51. For a different account of the necessity condition, see Seth Lazar, 'Necessity in Self-Defense and War', *Philosophy and Public Affairs* 40/1 (2012), 3–44.

[31] Blum, 'The Dispensable Lives of Soldiers', 69–124, argues that this wide permission is immoral; hence, the scope of combatant targetability must be legally narrowed. In her view, the decision

bombing military bases, killing enemy combatants, blowing up enemy tanks, and so on make at least some small contribution to the defeat of the enemy, they are always permissible regardless of their actual contribution in a given case. An attack (which inflicts no collateral harm on civilians) passes the *in bello* necessity test if its target is a military object.

Again, it is important to note that while the right to attack combatants is broad, it is not completely unrestricted. Although soldiers have a right to act on the presumption that killing enemy combatants promotes victory, state self-defence, and deterrence, if they happen to know that the presumption is false, the killing is impermissible. It follows that the killing must in some minimal sense be perceived as contributing to the defeat of the enemy, not as an act constituting what might equate with pure revenge. This limitation follows the contractarian logic: while the parties have an interest in exempting each other from the deliberative duty to verify that their attack is defensive, they have no interest in granting such an exemption in cases where such attacks are evidently gratuitous. Neither natural nor contractual morality licenses the killing of soldiers under such circumstances. Moreover, while combatants owe no duty to enemy combatants to make sure that attacking them contributes to the legitimate aims of the war, combatants ought to undertake efforts to ascertain that the attack is effective if such efforts are not risky and do not require special investment. While pre-contractual *in bello* necessity is relaxed by the agreement, good, professional soldiers would be guided by it under many circumstances, despite the liberty-right that they acquire to ignore it.

The contractual logic of *in bello* necessity explains both how *in bello* proportionality can be symmetrical, and how soldiers might be subject to this rule, despite the limited information that they have about the extent to which an attack may promote victory. We noted that decent states agree on an understanding of military value that does not require them to measure the extent to which an attack will contribute to victory. Accordingly, *in bello* proportionality does not require demonstrating causal connection between the contemplated attack and victory. Instead, for the purpose of the proportionality test, states agree to assess military value by mere quantity. Commanders are entitled to assume that since killing enemy combatants has military value, the more combatants killed, the greater the chances of victory by their side. Thus, since killing ten enemy combatants has greater military value than killing one, this killing justifies more collateral damage. The same is true of bombing bases, attacking convoys, and so on.

---

regarding whether or not to adopt the amendments she proposes 'must rest on the value we want to assign human lives, particularly those of enemy combatants' (75). In our view, the decision depends less on this value and much more on whether wars could be fought effectively with constraints of the sort Blum describes. Since fighting under the demand to distinguish 'threatening combatants from unthreatening ones' (108) would make effective self-defence too hard, the parties subscribe *ex ante* to a rule that exempts all combatants from it and defines enemy combatants as free game. As we shall explain shortly, by accepting the arrangement, soldiers acquire a liberty-right against their enemies, one that in many cases they should not exploit.

The proportionality rule does not require assessing the estimated contribution of the attack to victory against the estimated collateral harm to civilians. Instead, it weighs collateral damage and the immediate, concrete military value of the attack. The same is true of the killing of civilians: *in bello* proportionality does not distinguish between culpable and non-culpable civilians, or between collateral killing of civilians that promotes victory and collateral killing that does not.

When the military advantage is minor—eliminating only one rocket, or killing only one enemy combatant—the parties can assume that giving it up will not undermine their ability to attain effective self-defence and deterrence. True, at times attacks on minor military targets (one rocket or one missile, say) might be thought to promise a substantial gain (e.g. rescuing thirty children that this rocket would otherwise kill). Refraining from the attack on the rocket because it is predicted to kill ten civilians might be seen as too costly. But such predictions are usually unreliable. They are likely to stem from the parties' tendency to overestimate the value of the military objects they want to attack. Therefore, it is in the interest of the parties to the war contract to agree that attacks on objects of low military value should be aborted in case they are expected to cause excessive collateral harm to enemy civilians.

In one sense, then, *in bello* proportionality is much more permissive than pre-contractual proportionality: it allows the killing of innocent civilians in order to achieve an unjust cause. In another, it is much more restrictive: it prohibits the killing of culpable civilians even if this is necessary to achieve a just cause. Moreover, in many cases it rules out taking remote consequences into the proportionality calculation. One should not kill ten civilians while eliminating one small rocket, even if one believes that otherwise there is a good chance that this rocket will kill thirty children.

Our contractual characterization of military value resolves the conceptual concern about the moral equality of civilians; namely, the concern that killing just civilians as a side-effect can never satisfy the proportionality condition. According to contractarianism, when combatants deliberate about a possible attack that is expected to cause collateral harm, they may bracket two kinds of questions: first, whether the war they are fighting is just or not; and second, whether the proposed attack makes a significant causal contribution (or *any* such contribution) to victory.[32]

But what about cases in which the military value of such an attack is substantial? How many civilians are we allowed to kill in order to eliminate a huge military base? In our opinion, proportionality cannot offer a number or even a range of numbers. The incommutability of the goods (lives vs state self-defence, lives vs deterrence from future aggression) undermines any attempt to suggest a

---

[32] For the difficulty of defining victory, see Chapter 4 above.

definite answer. This has to do with the insurmountable epistemic difficulty in figuring out the right ratio between the military value of attacks *of the just side*, on the one hand, and the harm to civilians, on the other. The comparison is extremely difficult, especially because states *agree* to treat the territorial integrity of states and their independence as goods worth protecting by defensive war. The pre-contractual value of these goods is hard to compare to the value of the lives of innocent lives.

As shown in a recent study, even experts specializing in the ethics or in the laws of war cannot reach a reasonable consensus on what the proportionality test requires in practice.[33] More accurately, they agree only on the extremes, when the harm to civilians is either very low or exceedingly high. Since in the real world the numbers are usually somewhere between these poles, it seems that we cannot turn to experts in the field for reliable guidance.[34] Similar results emerged when the study was conducted among military personnel. Therefore, requiring warring parties to verify the due proportionality between military value and collateral harm before every attack involving (expected) harm to civilians would be impractical and would hinder their attempts to defeat their enemy. The parties release each other from such a requirement and undertake only a limited restriction; namely, to refrain from attacks whose military value is minor and that are expected to cause clearly excessive harm to civilians.

6.6 Civilian Acceptance of the War Agreement

We turn now to a different concern about the permission to inflict collateral damage on civilians. From a contractarian perspective, there seems to be a profound difference between civilians and soldiers. As argued above, soldiers willingly and knowingly join the army and accept their role.[35] Since soldiery is part of a mutually beneficial and fair arrangement, it can be assumed that they freely accepted their role, together with the rules that define it. This acceptance involves waiving moral rights and undertaking moral duties. They knowingly come to occupy their role and (perhaps unknowingly) accept the package of norms that constitute it. Acceptance is the locus of the moral efficacy of the war agreement: combatants have an equal right to kill each other in war because, by joining the army, they accept a regime under which they are released from the

---

[33] Daniel Statman, Raanan Sulitzeanu-Kenan Micha Mandel, Michael Skerker and Stephen De Wijze, 'Unreliable Protection: Proportionality Judgments in War' (unpublished manuscript).

[34] Or, if you wish, the study shows that there are actually no 'experts' in the relevant sense; that is, people who (thanks to their education or experience) can be trusted to provide better advice than others in some domain. For the view that endemic disagreement between moral philosophers undermines their claim to expertise in the fields of disagreement, see Ben Cross, 'Moral Philosophy, Moral Expertise, and the Argument from Disagreement', *Bioethics* 30 (2016), 188–94.

[35] See Section 5.2 above.

duty to verify that the war they are sent to fight is just. Similarly, all civilians are immune from intentional killing in war, since soldiers undertook a duty to avoid intentionally killing them, even if they are culpable.

However, there seems to be no similar act by which civilians could be said to accept being potential victims of *Collateral Damage*. *Civilians never undertake any specified civilian role*. Being a civilian is not one role among others that an individual might undertake (or refuse to undertake) together with the package of rights and duties associated with it. And if civilians cannot be said to have waived their right not to be attacked, combatants cannot be said to have been released from the duty to refrain from harming them, either intentionally or even as a side-effect. The consent of *combatants* to the *in bello* regime can thus make no difference to the moral standing of *civilians*. For those who believe that pre-contractual morality does not distinguish foreseeable collateral damage from intentional killing, this concern applies even to the collateral killing of Unjust Civilians.[36] However, it applies in a much stronger way to the collateral killing of Just Civilians, who never accepted a regime under which it would be permissible for Unjust Combatants to kill them collaterally.

This concern assumes that 'the only way in which we can acquire role obligations...with genuine moral force is by signing on for the roles to which they are attached'.[37] Yet, as the phenomenon of familial roles shows, one can occupy a role by being born into it. We did not choose to be sons/daughters or brothers/sisters, but nonetheless we have duties by virtue of these roles. Likewise, we do not choose to be citizens of the state in which we were born. Whereas actual choice is required in order to take on a specific role *within* civil society, it is not required for the sake of occupying the inherent role of a citizen. Nonetheless, it is a mistake to say that roles not chosen (like being a citizen or a daughter) are imposed on us *against* our will. They are non-voluntary, but they are usually not involuntary.

There are, of course, differences between familial relations and the relations constituted by political affiliation; but we believe these roles to have enough in common to justify the analogy. Note, in this context, the commonalities between the sentiment we have toward family members and the sentiment we have toward the state. One cannot be indifferent to the fact that one's brother is a murderer or rapist—shame is a natural emotion in such cases; likewise in the political context. Speaking of his experience as an American citizen during the Vietnam war, Nagel says, '[c]itizenship is a surprisingly strong bond, even for those of us whose patriotic feelings are weak. We read the newspaper every day with rage and

---

[36] For instance, Robert L. Holmes, *On War and Morality* (Princeton, NJ: Princeton University Press, 1989).
[37] Hardimon, 'Role Obligations', 343 (rejecting this idea).

6.6 CIVILIAN ACCEPTANCE OF THE WAR AGREEMENT 151

horror, and it was different from reading about the crimes of another country.'[38] Familial and national relations provoke positive emotions as well: the pride felt for the achievements of a compatriot resembles the pride felt for the achievements of a brother or cousin. This identification of civilians with their countries undermines the potential claim that since they never explicitly undertook the role of a civilian, it was forced upon them against their consent.

Now, most educated and informed civilians acknowledge the authority of states to act in their name and to represent their interests in the international realm. They might disagree or even resent some treaty-based arrangement to which states agree, but they acknowledge that there is no other way to take care of these interests. A regime in which states would be barred from undertaking commitments aimed at the advancement of national interests would be much worse, overall, for their citizens. We contend, in other words, that a mutually beneficial regime authorizes states to represent their citizens in these matters. By accepting a legal system that their state enforces, citizens authorize their state to act on their behalf in the international realm. Citizens habitually follow the social rules that empower decent states to enter such interstate arrangements. Hence, if fair and mutually beneficial, a treaty-based international law is as morally effective as the domestic law by which these individuals are governed.

This logic applies to the domain of war too. Legally, states under conditions of minimally just symmetrical anarchy may waive the legal right of their citizens not to be harmed collaterally, in exchange for a right to carry out military attacks in war. They do so in order to allow their armies to defend their contractual right against aggression by force, and to empower their soldiers to undertake the duty of obedience (assuming that obedient armies are necessary for efficient protection and enforcement of their contractual right against aggression). If *Collateral Damage* is *ex ante* mutually beneficial to decent parties in war, by undertaking it a state acts within the general power conferred on it by its citizens. In these cases, states are entitled to redistribute their citizens' pre-contractual rights so as to better protect their interests. It follows that civilians lose their right against collateral damage by accepting the rules that empower states to act on their behalf.

The moral status of civilians vis-à-vis their enemy combatants is basically no different from that of civilians vis-à-vis policemen who come to arrest them or prison guards who lock them up. In none of these cases is there an explicit undertaking of the detailed terms of the relevant social arrangements. The rules of these arrangements are nevertheless binding. True, in the absence of an explicit undertaking, civilians do not usually know exactly what their rights are vis-à-vis the role-holders mentioned above (policemen, prison guards, etc.). We have seen, though, that this is true even for those who explicitly undertake social roles. They

---

[38] Nagel, *Mortal Questions*, xii.

often have only a partial idea of what doing so means. Consider, for example, an illiterate combatant. Her illiteracy does not absolve her from the rules that apply to her qua combatant. In particular, illiteracy does not absolve her from being a legitimate target for an enemy attack, regardless of her actual contribution to the war effort and of her individual responsibility for the war fought by her country. Since the institution of soldiery is defined by the law, those who join it and become combatants are automatically subject to the rules that constitute this role. The same applies to civilians, despite the fact that it seems less natural to talk about the *institution* of citizenship and about *joining* it.

6.7 *Civilian Immunity* and *Moral Equality* in Asymmetric Wars

At the end of Chapter 3, we proposed that some non-state actors (NSAs) should also be considered parties to the war convention. Under specified conditions, they too might be justified in launching war.[39] In rare cases, nations fighting for liberation maintain proper armies, and in this case they might be subject to the regular *in bello* rules. Typically, however, such stateless nations do not have organized armies, and their ability to meet the enemy on the battlefield is very limited. Hence the asymmetry between the parties. The power differences in asymmetrical conflicts are so great that the traditional war contract is neither mutually beneficial nor fair; in a war of independence regulated by pre-contractual morality, the just weak side has better chances to achieve its just aim. Thus, the question is what rules apply to such actors once a war breaks out, and what rules apply to states fighting against them. The present section addresses these questions.

We begin by explaining why *Civilian Immunity* should not govern asymmetric wars, as this rule is traditionally understood. The *in bello* agreement assumes that in symmetric wars, the warring parties are expected to benefit from a rule that grants immunity to noncombatants from intentional attack. Such a rule would not decrease their chances of military victory, but it would reduce the horrors of war. This assumption is unrealistic with respect to asymmetric wars. Non-state actors in asymmetric conflicts might have no chance of achieving *military* victory, as military targets may be too far away or so well protected that the weak party cannot really harm them. Militants can effectively use force (if at all) to put psychological pressure on the enemy. The way to wield this pressure is not by weakening the army, however. Hence, the prohibition on targeting civilians 'seems hopelessly naïve particularly when a group cannot take on military targets because

---

[39] This section revisits and modifies the argument in Benbaji, 'Justice in Asymmetric Wars: A Contractarian Analysis', Sect. IV.

there is none in range, or they are too well protected'.⁴⁰ Effective attacks must be on civilians and civilian infrastructure, which will make life harder for the enemy civilians. They, in turn, will press their political leaders to negotiate with the NSA.

Under these circumstances, *Civilian Immunity* would work against the only goal that NSAs could achieve. That is, compared to a war governed by pre-contractual morality, warfare limited by *Civilian Immunity* would change the chances of victory. Such a prohibition would give a clear advantage to the strong side, even if it is an unjust colonial power that (violently) denies a national group its right to independence and self-determination. Such a rule would, therefore, violate *Fairness*: if the stateless nation has a just cause for war, *Civilian Immunity* would diminish its chances of promoting its goals by the use of force.⁴¹

Thus, applied to asymmetric conflicts, *Civilian Immunity* is unfair in two distinct senses. First, it might contribute to the solidification of existing unfair inequalities between powerful colonizers and weak colonized peoples. (Recall that the war agreement as a whole is unfair if respecting it solidifies such injustices in sufficiently many identifiable cases.) Second, fairness of the *in bello* contract requires that under the war agreement, the parties maintain the chances of victory that they would have had in an otherwise identical war that is regulated by pre-contractual morality. But *Civilian Immunity* eliminates the chance that weak, stateless nations have to promote their just or unjust aims by using force.

Which *in bello* code that meets *Fairness* can regulate asymmetric wars of independence? We will answer this question shortly, but first we note that the apparent unfairness of the *in bello* code in asymmetric wars extends to other aspects of *Civilian Immunity*. For instance, the duty to wear a uniform, which enables the parties in regular wars to maintain the distinction between combatants and noncombatants, hampers the ability of fighters in asymmetric wars to camouflage. If they obey it, they will make themselves easy targets for their enemies, practically paving the way to their defeat. Given the enormous military inferiority of NSAs, the only survival tactic for its militants is to hide among civilians (typically, 'their' civilians). Moreover, when the military objects of the strong side are unattainable, *Collateral Damage* offers no benefit to the weak side. The weak side cannot attack any military object and therefore cannot take advantage of the right to harm civilians collaterally while doing so. The striking result is that the strong side might kill many civilians legally, while its own civilians are legally protected.

We suggest that the *in bello* rule to which weak, decent national groups fighting for independence are subject grants them permission to attack civilian targets with

---

⁴⁰ See Michael Gross, *Moral Dilemmas of Modern War* (New York: Cambridge University Press, 2009), 199.
⁴¹ On the importance of the fairness condition in asymmetric wars, see also David Rodin, 'The Ethics of Asymmetric War', in Sorabji and Rodin (eds.), *The Ethics of War—Shared Problems in Different Traditions*, (Aldershot: Ashgate Publishing, 2005), 158–61.

the aim of causing fear and demoralization. They are allowed to harm civil society by targeting the institutions that constitute it, if this strategy might put enemy leaders under pressure to give up or to compromise. They are allowed to target civilian infrastructure, such as public buildings, monuments, roads, and so on. Since the *jus in bello* is independent of the *jus ad bellum*, this is true of those who claim to be freedom fighters but have no just cause for a war of independence.

Crucially, however, this licence does not permit direct attacks on civilians. Schools, buses, or old age homes should be left out. The rationale offered here is that attacks on civil society can often be carried out without harm to civilians by issuing a warning before the attack so that civilians can distance themselves from the attacked target. When such warning takes place, the attacks are permissible.[42]

But would advance warning not diminish the effectiveness of such attacks? We believe not. In most cases, the ability of the attacks on civil society to influence public opinion does not depend on killing civilians, but on exposing the vulnerability of civil society. The expected benefit to the NSA is attained by the actual attack on public space—government buildings, power stations, and central tourist sites—even when no civilians are harmed (having followed the warning and distanced themselves from the planned target). In fact, killing civilians often has an adverse effect. It creates loathing, which makes public opinion more obdurate, thus functioning against the political goal of the organization.[43]

Moreover, it is unclear whether it is useful to target politicians, who are quite clearly culpable for the assumed injustice meted against the NSA. First, doing so often leads to retaliation and to a slighter chance of achieving the required political goals. Second, removing politicians is risky because even worse ones might replace them. Hence, we may presume that in general, a prohibition against targeting culpable enemy civilians would not disadvantage the weak party in the conflict. In some rare cases, assassination might be useful, but given the proneness to err and the danger of escalation, there is a clear interest in granting immunity to all enemy civilians.

We thus suggest that traditional *Civilian Immunity* that prohibits targeting civilian objects systematically discriminates against those who fight just or unjust wars of independence. In contrast, the rule that allows targeting civil society but grants immunity to civilians is mutually beneficial and fair; it is the rule to which just and unjust NSAs are subject, when fighting asymmetric wars of independence.

---

[42] For the practice and legal status of such pre-warning, see Pnina Sharvit-Baruch and Noam Neumam, 'Warning Civilians Prior to Attack under International Law: Theory and Practice', *International Law Studies* 87 (2011), 359–412.

[43] For the claim that terrorism tends to fail for this and other reasons, see, for example, Max Abrahms, 'Does Terrorism Really Work? Evolution in the Conventional Wisdom since 9/11', *Defense and Peace Economics* 22 (2011), 583–94.

The question that naturally arises at this stage is whether the release of NSAs from the prohibition against attacking civilian targets entails a parallel release for the states fighting against them. At first glance, we might assume that such a release is indeed entailed. Let us explain why. To secure a high level of compliance with the rules of war—and with legal rules in general—rule-makers must opt for rules that embody formal equality. Indeed, most international law is articulated in uniform and generic terms making no distinction in obligation or in permission for differently situated parties. Rules that do not express such equality are harder to apply because they require prior judgment about which rules apply to each player.[44] They could also make the strong side feel that it is put at a disadvantage—facing attacks on its civilian buildings, facilities and institutions—without being allowed to strike back in kind. This could lead it to (unjustifiably) violate the rules that apply to it under the current proposal. Therefore, if the contract allows NSAs to attack civilian targets after proper warning, it must allow the same course of action to the states fighting against them.

This is a powerful argument, but the opposite one seems stronger. The permission for NSAs to attack civilian targets is radically constrained. This constraint provides an incentive for states to enter into this agreement. *Ex ante*, the part of the war agreement that applies to asymmetrical wars is, therefore, mutually beneficial to both sides. If states benefitted from the same permission to target civilian infrastructure that is granted to NSAs, the unfairness this arrangement attempts to overcome would re-emerge. States could then use their military might to ruin the civil life of their enemy by systematically bombing its infrastructure (after proper warning). They could do so in a much more devastating manner than NSAs could, given the latter's limited military capacities. Indeed, the historical record on insurgency and counterinsurgency indicates that in the absence of censorship at home, counterinsurgency tends to lead to extreme brutality. While the permission granted to the weak side to target civil society does not endanger the very existence of civil life in the country, a parallel permission to the strong side does carry this danger.

States may still consider themselves disadvantaged by this seemingly asymmetric regime that grants their enemies permission not available to them. They may assume that if they are barred from carrying out similar attacks on civilian infrastructure, they will be left with no effective way of defending themselves from painful attacks on their towns, national symbols, public buildings, and so on. In the absence of standard military targets—headquarters, bases, training camps, and brigades preparing for battle—and assuming the regular *in bello* restrictions on the use of military force in war, the state's jets, tanks, and submarines would be

---

[44] See Gabriella Blum, 'On a Differential Law of War', *Harvard International Law Journal* 52/1 (2011), 163–218.

of no use. The complaint, then, is that in an attempt to remedy the unfairness against NSAs, the war agreement would be creating unfairness against the *states*.

However, this complaint is unwarranted. The military superiority of states over NSAs affords states many ways of fighting NSAs effectively without having to resort to attacks on civilian infrastructure. Given their intelligence and their technological capabilities, states are usually able to trace and target their enemy's militants and the military facilities, as well as to improve protection over their own civil life and infrastructure. The ability of NSAs to cause real harm to the civil life of the states they fight against is much more limited—at least insofar as they follow the *in bello* rules against intentional attacks on civilians and the corollary rule mandating the issue of warnings before attacks on civilian targets. When NSAs comply with these rules, there seems to be no unfairness in the fact that states are subject to a more restrictive rule regarding attacks against civil society.

We emphasize that surely, those NSAs that fight for national liberation by indiscriminate attacks on civilians, or that target civil society without pre-warning civilians, constitute a different category. States who suffer from a violation of the contract must have remedial rights. We construct the remedial aspect of the war convention in the next chapter.

Let us connect this discussion on the *in bello* rules that regulate asymmetric wars to our discussion of independence wars in Chapter 3. The expression 'symmetrical anarchy' was intended to capture two aspects of the pre-contractual state of affairs on the international plain: (a) the anarchy aspect—the absence of a common power that could resolve conflicts between parties and effectively enforce the resolution; and (b) the assumption that it is *ex ante* beneficial to all states to enter an agreement that prohibits wars whose aim is the enforcement of pre-contractual justice. In Chapter 3, we further assumed that even if the power inequalities are huge (the chances of the weak side to win a war governed by pre-contractual or contractual morality are very low indeed), the use of force is usually suboptimal to *both* sides. It would be better for both sides to bargain rather than to fight. Thus, the 2(4) prohibition and the 51 permission apply to wars of independence: the weak side may use force only if the strong side uses force. Stateless nations, we suggested in Chapter 3, are subject to the general prohibition on the use of force, just like states under conditions of minimally just symmetrical anarchy. The Charter's *jus ad bellum* rules should be extended to conflicts between states and stateless nations that struggle for independence. The *ad bellum* prohibition on use of force is fair in such contexts as well.

By contrast, in this section, we argue that applied to wars of independence, the symmetric *in bello* contract is unfair. Under a symmetrical anarchy, a symmetrical *in bello* contract meets the conditions of *Mutual Benefit* and *Fairness*. But this code is designed to regulate wars between national armies of decent states. A stateless nation might have no army, and a regime that combines *Civilian*

Immunity and *Collateral Damage* eliminates the chance of the weak side to harm civil society even in cases in which this is the only way to address the aggression of the strong side. Therefore, we argued that the *in bello* regime to which stateless nations are subject should be interpreted differently. The weak side is allowed to target civil society, while taking all feasible measures to spare civilians, whereas the strong side is subject to the traditional rules of *Civilian Immunity* and *Collateral Damage*.

Finally, consider again the suspicion we raised in Chapter 3 regarding the unfairness of the war agreement as a whole towards stateless nations that have no recognized territorial rights nor regular armies. One might argue that one manifestation of this unfairness is the refusal to acknowledge the belligerent status of non-state fighters. Although the Additional Protocol I to the Geneva Convention asserts that 'those fighting [against colonial domination and alien occupation and racist regimes who] obtain prisoner-of-war status if captured, and immunity from prosecution from belligerent acts', some states, most notably the United States, object to this piece of legislation, arguing that it legitimizes terrorism. To avoid this charge, the Protocol created two classes of combatants: 'those fighting [against colonial domination and alien occupation and racist regimes who] obtain prisoner-of-war status if captured, and immunity from prosecution from belligerent acts'; and 'those fighting for less favored political causes... [who] would not receive POW status or immunity of prosecution from warlike acts'.[45]

Our analysis offers a more radical interpretation of the *in bello* agreement. Since the *in bello* code and the *ad bellum* one are independent of each other, militants ought to be treated as soldiers in the legal sense even if they have no just cause for their war of independence or if they unjustifiably resorted to war. Needless to say, if they violate the *in bello* rules, for instance by targeting civilians, they will be treated as criminals—just as would regular soldiers who do so. While the *in bello* code allows fighters of all organized armies or militia to kill each other (provided that these organizations can be seen as acting on behalf of a country or a people), it does not allow them to target civilians. Moreover, if our interpretation of the *in bello* code is right, militants who do whatever they can to minimize civilian casualties but spread terror by sabotaging civil society are entitled to the protections the LOAC provides to soldiers.

---

[45] Abraham Sofaer, 'Agora, The US Decision not to Ratify Protocol I to the Geneva Conventions on the Protection of War Victims (con'd), The Rationale for the United States Decision', *American Journal of International Law* 82 (1987), 785–6. Quoted in Michael Gross, *Moral Dilemmas in Modern War: Torture, Assassination, and Blackmail in an Age of Asymmetric Conflict* (New York: Cambridge University Press, 2009), 207.

6.8 Insights from Behavioural Ethics

In the last two decades or so, a new field has emerged in psychology—the field of behavioural ethics—that 'seeks to understand how people actually behave when confronted with ethical dilemmas'.[46] We believe that the insights coming out of this field support our central line of argument. In this section we briefly show how.

Let us start with the distinction between the 'should self' and the 'want self'.[47] Our should self 'is rational, cognitive, thoughtful and cool-headed', while the want self is 'emotional, affective, impulsive, and hot-headed'.[48] Our should self includes our beliefs about what we ethically ought to do; our want self reflects our actual behaviour. Research has shown that while the should self dominates our thinking before and after we make a decision, it is the want self that often wins at the moment of decision.[49] If we reflect calmly on whether we are allowed to lie for the sake of profiting, we easily, almost intuitively, know that doing so is unethical, and we predict that we would resist temptation to make such a profit if ever faced with such a test. But then, when we are actually confronted with such circumstances, we often fail and follow the guidance of our self-interested want self.[50] After the event, our should self dominates again, either causing us to 'forget' what we actually did or rationalizing it and thereby maintaining our self-image as people of moral integrity.

The claim is not that people always fail to respect their moral obligations (more accurately, what they themselves identify as their obligations in moments of reflection); but that given certain conditions, they often do. Our ability to transcend our self-interested nature is much more limited than we would like to think, especially when basic interests seem to be at stake:

> At the time of the decision, visceral responses lead us to an inward focus dominated by short-term gains. We direct our attention toward satisfying our innate needs, which are driven by self-preservation. Other goals, such as concern for others' interests and even our own long-term interests, vanish.[51]

This inward focus and the ethical failures to which it leads are intensified, especially at the organizational level, by uncertainty and time pressure. Studies

---

[46] Bazerman and Tenbrunsel, *Blind Spots*, 4.
[47] See Max H. Bazerman, Ann E. Tenbrunsel, and Kimberly Wade-Benzoni, 'Negotiating with Yourself and Losing: Making Decisions with Competing Internal Preference', *Academy of Management Review* 23 (1998), 225–41.
[48] Bazerman and Tenbrunsel, *Blind Spots*, 66. [49] Ibid.
[50] For research showing that most decent people would lie (though only to an extent) if they think they can get away with it, see Dan Ariely, *The (Honest) Truth About Dishonesty: How We Lie to Everyone—Especially Ourselves* (New York: HarperCollins, 2012).
[51] Bazerman and Tenbrunsel, *Blind Spots*, 71.

have shown that the more uncertain the environment, the greater the time pressure agents experience, the more likely they are to engage in unethical behaviour.[52]

A possible response to this growing body of research would be that although psychologically interesting, it is of no interest to philosophy, in particular to moral and political philosophy. The frequent failure of people to carry out (what they regard as) their moral obligation is not news. Why should it affect our understanding of what these obligations are, or of how ethical theory should be construed?

A central assumption of contractarianism is that morally effective legal rules should not be a simple copy of truths about pre-contractual morality. They must be easily recognizable, hard to deny, and easily enforceable, in a way that would reduce the gap between the should self and the want self. In other words, behavioural ethics does not simply teach us the depressing lesson that human beings are weak-willed and too frail to comply with the dictates of morality; it also clarifies the nature of the obstacles for such compliance and, consequently, points to the ways by which legal systems help us to overcome them.

These considerations are crucially important to the ethics of war, especially at the *in bello* level. When soldiers are under fire, when they see their comrades being hurt, their 'visceral responses' lead them to the focus mentioned above on their innate needs, which are driven by self-preservation. Compounding these conditions are the incurable uncertainty ('the fog of war'), time pressures, and frequent orders issued by their superiors, thus expanding even further the decision-making difficulties facing soldiers. Given these facts, it is impractical to expect soldiers to make subtle distinctions between, for instance, liable and non-liable targets, or to assess the military benefit (in terms of advancing victory) of attacking some target. Usually, combatants lack the time or the resources to conduct such inquiries and, in any case, they would be biased in carrying them out.[53]

This reality invites a service conception of (at least some) laws of war. If soldiers were required to make judgments about liability and causality in accordance with pre-contractual morality, that would lead to a rather weak protection of human rights. Hence, morality itself requires that partial, limited subjects such as human beings be placed under the authority of legal rules that will help them to comply with their moral commitments. More specifically, we suggest that states should accept *Civilian Immunity* because it has a legitimate authority over most soldiers

---

[52] Ibid., 164. See also Laura L. Nash, *Good Intentions Aside: A Manager's Guide to Resolving Ethical Problems* (Boston, MA: Harvard Business School Press, 1993), 166: 'Short-term pressures can silence moral reasoning by simply giving it no space. The tighter a manager's agenda is, the less time for contemplating complex, time-consuming, unpragmatic issues like ethics.'

[53] In a similar vein, see Larry May, *War Crimes and Just War* (New York: Cambridge University Press, 2007), 231: 'The rules of war constitute a system of norms for regulating the behavior of states and their agents in the absence of a World State. And the system of norms is meant to apply to probably the most stressful of times... In such times, any agreement about what the rules of the game are must be seen as a good thing.'

under most war circumstances. Moreover, if most soldiers followed these *in bello* rules most of the time, innocents would enjoy better protection in war. Hence, the parties should agree upon rules that only roughly and partially correspond with those issued by pre-contractarian morality, but have a much better chance of being followed. For instance, the message to fighting forces is that noncombatants are unconditionally out of the game except for the rare cases in which they take a direct or active part in hostilities.[54]

If such rules are accepted, combatants are released from the need to investigate whether any specific civilian comprises a legitimate target due to his or her moral responsibility for the seemingly unjust war and the expected military benefit from killing him or her. By the same token, as we saw in the previous chapter, combatants are exempted from the requirement to make sure that the soldiers they attack are legitimate targets according to the principles of pre-contractarian morality. For the most part, the distinction between these two groups is pretty clear; hence, it will be hard for combatants to ignore it, deceive themselves about its implications, or deny its moral significance.[55] This is why a rule incorporating this differentiation would be the one accepted by partial and decent states.

As we understand it, contractarianism takes the gap between the should self and the want self very seriously, especially in the face of grave threats to individuals or to collectives. According to contractarianism, morality handles this gap by requiring states to subject their soldiers to rules that are easily recognizable and easily enforceable. The warring parties would be better able to follow these rules when they confront serious threats to national security or in the midst of the turmoil of war.[56]

Some readers might feel that this approach confuses true morality with practical considerations. Being guided by such considerations is understandable, they might say, but one ought not to confuse that with *morality*. This response reflects the problematic distinction we mentioned earlier between the deep morality of war and the optimal laws of war that parties habitually obey. We address this issue once again in the concluding chapter of this book.

---

[54] For a proposal regarding how this condition should be understood, see Nils Melzer, *Direct Participation in Hostilities under International Humanitarian Law* (Geneva: ICRC, 2009).

[55] It would be hard, but of course, not be impossible. As Alex J. Bellamy shows at length in *Massacres and Morality: Mass Atrocities in an Age of Civilian Immunity* (New York: Oxford University Press, 2012), the rule of civilian immunity is 'relatively recent and fragile', and 'although it has become progressively more difficult to get away with mass atrocities, certain contexts continue to make that a very real possibility for some' (14).

[56] Of course, we do not assume the possibility of rules that are 100 per cent clear in their application with no grey area at all. Rather, we claim that the relevant rules should be crafted in a way that makes them as easily identifiable and applicable as possible.

6.9 Conclusion

Keeping civilians 'out of the game' in times of war is widely accepted as the most fundamental principle in the ethics of warfare. Harming them is permissible only as a side-effect and only if the harm is not disproportionate to the value of the military target. In terms of pre-contractual morality, however, this principle is difficult to justify. It is too restrictive in its blanket prohibition on the intentional killing of enemy civilians, and too permissive in at least two respects: (a) the right it grants Unjust Combatants to harm Just Noncombatants collaterally, and (b) the right it grants Just Combatants to inflict collateral harm on Unjust Noncombatants without establishing a causal connection between the military gain that justified such harm and military victory.

We claimed in this chapter that if the war agreement regarding the role of armies is universally accepted, then a further agreement that makes all civilians immune from intentional attack *ex ante* is mutually beneficial to all parties. By accepting it, states spare their civilians without reducing their chances of victory. Although in rare cases, attacking civilians does contribute to victory, decent parties that allow each other to defend themselves by instituting armies have a common interest in agreeing on a rule that prevents them from targeting any civilians, including culpable ones. Similarly, a rule that allows collateral damage to civilians is mutually beneficial because if such damage were ruled out, it would be much harder for armies to defend their states effectively.

Similar considerations explain the contractual *in bello* proportionality that prohibits the achievement of military advantage by inflicting excessive collateral damage to civilians. The parties are exempt from the need to establish the magnitude of the causal contribution of their attack to victory and to compare it to the expected harm to civilians. The exemption granted to combatants in this context has two related aspects. Combatants are exempted from making sure that their war is just and from establishing the causal connection between their attacks on military targets and victory. Nonetheless, while they do not *wrong* their enemy if they pointlessly weaken its military forces, they morally ought not to do so if they can easily ascertain that the attack would be either ineffective or unnecessary for victory.

In the terms taken from the literature on individual self-defence, the contractarian understanding of *Moral Equality* and of *Civilian Immunity* implies that combatants have a right to treat enemy combatants *as if* they were all innocent aggressors, hence legitimate targets for defensive attack (under a Thomsonian conception of self-defence). The same literature implies a duty for combatants to treat enemy noncombatants as if they were all innocent bystanders who may not be targeted even if doing so would benefit the potential victim.

Finally, we argued that applying the traditional *in bello* rules to NSAs (fighting against states) might be unfair to them. Typically, such actors do not have the military capabilities required to confront regular armies in the battlefield. If they were prevented from attacking civil society, they would be left with no effective way to defend themselves against injustices meted out to them. We suggested that NSAs should be released from the blanket prohibition against attacking civilian targets provided that they do not target civilians and that they attempt to reduce collateral harm to civilians. They may attack civilian *infrastructure*—public buildings, bridges, road junctions—but they must pre-warn civilians who might be harmed. This interpretation of the *in bello* code presumes that, typically, such caution in the targeting of civilian life would better advance their cause than would indiscriminate attacks.

States as well as NSAs sometimes believe that the legal and ethical constraints on fighting put them at a disadvantage in their attempts to defend their interests. They occasionally assume that complying with these constraints is a luxury they cannot afford. Consequently, they violate the *in bello* rules; they conduct indiscriminate attacks on residential areas, use illegal weapons, murder POWs, and so on. This behaviour seems to undermine the war contract altogether and to bring the parties back to the state of nature. Whether one-sided violation actually merits reversal to a state of nature, and what responses are acceptable in face of such violations, will be the topic of our next chapter.

What if combatants know that certain civilians are culpable for the unjust threat against which they are fighting and that targeting them is going to significantly contribute to victory? Should they target these civilians despite the contractual duty not to do so? One might assume that the answer is positive; after all, contractarianism implies that if combatants know that their war is unjust, they ought to refuse to participate in it, despite the liberty-right that they acquire to participate in the war. The analogy seems straightforward: combatants ought to kill civilians if they know that these civilians are culpable. However, the cases are substantially different. By accepting *Civilian Immunity*, culpable civilians acquire a claim-right against being targeted. Hence, even in the exceptional cases in which targeting such civilians promotes victory, Just Combatants violate their right against being targeted.

7
When the Agreement Collapses

7.1 Introduction

The preceding chapters showed that the logic that underlies social arrangements in decent societies applies within the context of war as well. We showed that under conditions of minimally just symmetrical anarchy, wars between decent parties that are governed by the traditional war convention are less harmful and involve fewer violations of rights than wars governed by pre-contractual morality. The theory proposed in Chapter 2 implies that if the war agreement is actually followed within a minimally just international community, it is morally effective.

The purpose of the present chapter is to address a major question raised by this argument. How ought the parties to respond if one of them deliberately ignores the rules of *jus in bello*—bombs civilian targets, murders POWs, and so on? If the ultimate moral basis for the war convention is contractarian—if the war convention is morally effective because all sides accept it—what moral restrictions would apply to Party B in case Party A manifestly rejected the contract?[1]

Minor violations do not usually constitute a reason to abandon a contract. They are pervasive and, to an extent, predictable and unavoidable. Often they result from the inability of the sides to predict the future and to assess the exact significance of all the details upon which they agree. For these reasons, upholding a contract that has been broken in a non-essential way usually remains mutually beneficial to both sides, especially where certain remedial rights are conferred on the disadvantaged party. Hence, negligible violations of the *in bello* rules, or isolated incidents that violate them, do not interest us here. We are concerned with cases where a systematic, fundamental violation of these rules occurs—a violation that indicates an abandonment of the contract.

One might assume that such violation of the war convention by one of the parties invalidates the convention, which presumably means that the parties revert to being governed only by pre-contractual morality. Pre-contractual morality may seem to be the default position to falls back on in case social arrangements that entail a different division of rights and duties collapse. Yet this response should be resisted. When the parties to the negotiating table agree on rules to govern the

---

[1] For earlier discussions of this question, see, e.g., Mavrodes, 'Conventions and the Morality of War', 128; Douglas Lackey, *The Ethics of War and Peace* (Upper Saddle River, NJ: Prentice-Hall, 1989), 62; and Waldron, *Torture, Terror and Trade-Offs*, 99, 105.

practice of war, they already take into consideration the possibility of rogue states, and they shape the contract accordingly. Therefore, we posit that the war contract includes rules that enable decent parties to deter other decent parties from becoming indecent or rogue, and, relatedly, rules about what may and what may not be done to those who violate the *in bello* rules. As we noted in Chapter 2, the war agreement is mutually beneficial only if it includes rules about situations involving some rejection of its rules and non-compliance with it. Arguably, such enforcement rules are inherent to the contract. We thus suggest that deciphering what may be done if the enemy abandons an important aspect of the war contract depends on the answers to two questions: First, to which rules must decent parties agree in order to deter decent states or stateless nations from such a violation? Second, which rules could parties undertake to deter decent parties from becoming indecent?

We start by articulating the rules that enable states to prevent the violation of the *in bello* agreement by 'someone who has a moral conscience and accepts the right rules' yet *might* violate them, because he or she 'does not care enough about morality to ensure good behaviour'.[2] These rules have two roles. First, they confer remedial rights whose very existence is supposed to deter decent parties from violating the *in bello* agreement. Second, exercising these remedial rights (by harming those who fail to fulfil their contractual duties) is expected to convince decent violators to re-commit themselves to the *in bello* agreement. To achieve these ends, the war agreement confers a right of retaliation.

We then turn to those terms in the war agreement that are designed to prevent states from becoming rogue or indecent and to the remedial rights that decent states have if they are threatened by rogue states. Rogue states are led by politicians who are unmitigated amoralists; viz., people whose political deliberation is insensitive to moral considerations. As we read it, the agreement confers on decent states a right to hold weapons of mass destruction (WMD) if this is necessary to deter other decent states from becoming rogue. Thereby, the agreement allows imposition of a direct threat on the people on whose behalf rogue states act, providing an incentive to these people to prevent such evil individuals from becoming their leaders. The agreement further includes a remedial right to use these weapons (i.e. to launch indiscriminate attacks) on a rogue enemy if the massive massacre that the rogue enemy threatened comes to maturity.

The war convention's assumption of this structure is evident at the *ad bellum* level, where the focus of the agreement is on what may be done to parties who violate the (contractarian) rule against first use of force. Recall that the right granted to states to wage war against violators of their territorial integrity is not a result of the collapse of the war contract—of a withdrawal, so to speak, to the state

---

[2] Hooker, *Ideal Code, Real World: A Rule-Consequentialist Theory of Morality*, 82.

of nature. Rather, it is a term in the contract itself. The victim of the aggression may address it by weakening the aggressor's army in a way that blocks the aggression, restores any violation of its territorial integrity, and deters the aggressor from further violations of the prohibition in the future. We suggest that the same applies at the *in bello* level. In contrast to the impression one might get from the title of this chapter, the war contract never really 'collapses'.

In Section 7.2, we explore the *in bello* rules to which the parties are subject in cases where one of them systematically violates these rules. In Sections 7.3 and 7.4, we deal with threats of an especially high magnitude that traditionally fall under the title of 'supreme emergencies'.

7.2 Ruthless Warfare and the War Agreement

Assume that war breaks out between State A and State B as a result of a dispute about some territory to which both states have claims. Soon after the outbreak of war, it becomes clear that State A has very little respect for the conventional *in bello* rules. It deliberately attacks residential areas in the capital city of State B, it uses weapons that are forbidden by the Geneva Convention, it kills POWs, and so on. What does such systematic violation of the *in bello* code imply with regard to the permissions and obligations that apply to combatants of State B?

To begin answering this question, let us first note a distinction to which we have already appealed. Consider rules whose justification seems merely conventional—for example, the prohibition on shooting pilots bailing out of crippled aircraft.[3] (Indeed, no similar rule exists to restrict the shooting of combatants attempting to escape from landcraft, e.g. from burning tanks.) Or consider rules prohibiting the use of certain kinds of weapons, for example, chemical weapons (used, let's stipulate, only against combatants), or laser weapons designed to cause blindness.[4] The basis for these rules seems solely conventional. If all sides refrain from using chemical weapons, the horror of war will be reduced without blocking the possibility of effective self-defence and without giving one side an unfair advantage over another. Pre-contractually, however, *if* it is justified to kill some person in war or to cause severe injury and pain to her, say, by firing a missile at her tank and burning her, it seems arbitrary to forbid the use of these latter measures, assuming such measures pass the *in bello* necessity test.

The mere conventionality of these rules is manifested in the fact that they become null and void in the face of their systematic violation. In cases like the immunity granted to pilots parachuting from airplanes or the prohibition on the

---

[3] See Article 42 of the Protocol Addition to the Geneva Convention.
[4] See Protocol IV of the 1980 Convention on Certain Conventional Weapons, issued by the United Nations in 1995.

use of some weapons, violations by one side are sufficient to exempt the other from observing such rules.

But this way of addressing the violation of the agreement cannot be generalized. First, there are prohibitions whose validity seems to survive their violation by the enemy. If your enemy bombs civilian hospitals or kindergartens, this provides you with no moral justification to do the same to their sick and young children. Moreover, a systematic violation of the rules by one side does not imply that the conflict is now governed by asymmetrical pre-contractual morality. Had it been so, Just Combatants would be barred from killing innocent Unjust Combatants in case culpable Unjust Combatants systematically violated the *in bello* contract. But such a scenario seems hard to believe; *Moral Equality* is not undermined in face of such a violation. Just Combatants do not lose their right—their *contractarian* right—to use military force against innocent Unjust Combatants even if other Unjust Combatants systematically violate many of the *in bello* rules.

Hence, we need a different conceptualization of the 'collapse' of the *in bello* contract. Consider a minimally just symmetrical anarchy in which *all* parties are decent (rather than merely most of them). Under such circumstances, maintaining the war convention is beneficial to all parties. States are under moral duty to enter the *in bello* agreement, as it improves the protection of (pre-contractual) human rights.[5] Still, when states enter this arrangement, they ought to devise ways to ensure that the contract is respected. The reason why such incentives are particularly necessary is that we all know *ex ante* that the temptation to breach the contract *ex post* will be very strong. It is one thing to agree *ex ante*, on the basis of mutuality, not to deliberately kill civilians or not to use weapons of mass destruction. It is a different thing to actually stick to these commitments when violating them is strongly believed to improve the chances of victory against an aggressor. Thus, contractarianism takes very seriously the possibility that states and NSAs would systematically violate the rules of war.

We suggest that the war agreement includes rules whose aim is to deter parties from violating the contract and to compensate the side that suffered from such a violation. These rules confer remedial rights to respond in one way or another to such violations. Thus, in effect, it is not pre-contractual morality that determines the right to use laser weapons in response to their use by the enemy, but rather the war agreement itself. When the *in bello* rules are violated systematically, the contract permits reactions that otherwise would be (contractually) forbidden, with the purpose of stopping the violation. Unsurprisingly, this is how contemporary international law understands the legal right to carry out reprisals. As Eyal Benvenisti puts it:

---

[5] This section revisits and modifies the argument made by Yitzhak Benbaji, 'Contractarianism and Emergency', in Hanoch Sheinman (ed.), *Understanding Promises and Agreements: Philosophical Essays* (New York: Oxford University Press, 2010), 342–65.

[T]he logic of reciprocity stipulates that the violation of rules by one side entitles the opponent to respond in kind. This logic has spawned two doctrines in international law. Under the first, the violation of the laws of war by one side releases the other from the obligation to follow the law. Under the second, the violation entitles the opponent to resort to reprisals—counter violations in kind. *Contemporary governments still adhere to this second doctrine.*[6]

The contractarian framework also explains why the response usually should be in kind, which is hard to comprehend if, as a result of systematic violations, the parties were under the authority of pre-contractual morality. It does make sense, however, if the licensed response is seen as a remedy offered by the contract itself.[7] The response in kind is supposed to deter potential decent violators of the *in bello* rules. It is further supposed to help violators re-gain their wit and realize that, notwithstanding appearances, the *in bello* rules—such as the prohibition against certain weapons and that against shooting pilots—are beneficial to all sides. The reprisal should convey the message that the retaliator is ready to re-enter the contract, if the violator accepts it. This signal is clearer and more powerful when the counter-violation is in kind than when it concerns some other aspect of the war agreement.[8]

Things get less clear when we turn to (violations of) *Civilian Immunity*. Which remedial right is created when the enemy systematically ignores the duty not to attack civilians intentionally? An in-kind remedial right seems out of the question. Supreme emergencies aside (see below), the moral foundations of the *in bello* contract will not tolerate the intentional killing of children. The reason is that *Civilian Immunity* is at the core of the war agreement in the sense that states are under a pre-contractual obligation to enter into an agreement that prohibits

---

[6] See Eyal Benvenisti, 'Human Dignity in Combat: The Duty to Spare Enemy Civilians', Israel Law Review 39/2 (2006), 103emphasis added. See also fn. 84 in Benvenisti's paper, for references to the objections raised by the United States, France, and the UK to an unqualified prohibition against retaliation. In ratifying the first additional protocol of 1977, in 1998 the United Kingdom submitted an illustrative reservation: 'If an adverse party makes serious and deliberate attacks, in violation of Article 51 or Article 52 against the civilian population or civilians or against civilian objects, or, in violation of Articles 53, 54 and 55, on objects or items protected by those Articles, the United Kingdom will regard itself as entitled to take measures otherwise prohibited by the Articles in question to the extent that it considers such measures necessary for the sole purpose of compelling the adverse party to cease committing violations under those Articles, but only after formal warning to the adverse party requiring cessation of the violations has been disregarded and then only after a decision taken at the highest level of government.'

[7] For the idea that reprisals 'have as their purpose the enforcement of the war convention', see Walzer, *Just and Unjust Wars*, 209.

[8] We agree with McMahan that 'whether a party that has thus far complied with a conventional law of war has a reason, moral or legal, to continue to comply when another party has begun to violate that law may depend simply on whether the violator is more likely to be motivated to resume compliance by reprisals in kind or by a show of good faith in continued adherence by its adversary'. See McMahan, 'The Morality of War and the Law of War', 35.

killing civilians in order to spare the innocents. Once decent parties are subject to the war agreement, they are under a contractual requirement to retain *Civilian Immunity* in the face of its violation by the enemy, thereby (a) re-affirming their commitment to the war contract and (b) signalling to the enemy that the road back to the *in bello* contract is still open.

If not an in-kind response, which type of remedial right is generated by a systematic violation of *Civilian Immunity*? Consider civilians who are morally responsible for the perceivably unjust war fought by their country and causally contribute to it, such as politicians, administrators, officials, military scientists, and so on. Killing them would presumably deter the enemy from continuing further violation of *Civilian Immunity*. In terms of pre-contractual morality, the standing of these civilians is no different than that of many enemy combatants. Hence, we suggest that if one side breaches *Civilian Immunity*, the other side is released from the contractual prohibition on killing involved civilians. One might be justified in responding somewhat in kind by targeting politicians, administrators, and other *involved* civilians on the enemy side; or similarly, with attacks on civilian infrastructure such as power facilities, ports, bridges, highways, and so on. As we noted in the previous chapter, these actions presumably do not promote victory. Yet when one (generally) decent party abandons the contract, retaliation and reprisals in kind might be necessary in order to restore compliance. The decent party is not released from the demand not to attack uninvolved, truly innocent enemy civilians. The purpose of violating *Civilian Immunity* in response to a violation by the enemy must be to force the violator to respect the contractual immunity of civilians. This enforcement must be carried out cautiously and wisely, sending a clear message about the intolerability of the violations committed, on the one hand, while, on the other, leaving the door open for the violator to renew commitment to the contract.[9]

The permission to target civilians who contribute directly to the war effort is curtailed in two ways: First, only essential breaches bring about a normative transformation of the kind described here. Notwithstanding the common rhetoric in times of war, the enemy state (or enemy NSA) cannot be said to have abandoned its contractarian obligations just because some of its military units misbehaved, either maliciously, or due—more typically—to factual or normative mistakes. Second, as emphasized above, the 'taking off of one's gloves' must be

---

[9] Douglas Lackey, followed by Stephen Nathanson, reject the contractarian argument for noncombatant immunity, claiming that if this rule were merely conventional one would be released from the duty to respect it in case one's enemy didn't, which seems to them outrageous. Since under our approach, some cases of killing civilians are only contractually forbidden, we are willing to bite the bullet. At any rate, contrary to their understanding, contractarianism is not committed to the view that when others violate the *in bello* rules, one's obligation to respect them automatically terminates. See Lackey, *The Ethics of War and Peace*, 61–2, and Stephen Nathanson, *Terrorism and the Ethics of War* (New York: Cambridge University Press, 2010), 237–8.

extremely limited. Violations of the contract do not provide states with a licence to free themselves from all contractual obligations towards their enemy.[10]

Politicians of Western democracies tend to regard all non-state organizations that use force against them as 'terrorists'. They believe that militants require a different moral and legal treatment than do soldiers in proper armies. In previous chapters, we rejected this view. Insofar as non-state actors (NSAs) respect the war convention, mainly by refraining from targeting civilians, they should benefit from the same moral rights as regular armies.

Often, however, such organizations do not even try to spare civilians. Consider an organization like ISIS, whose goals and methods are contrary to all accepted rules regarding war, both on the *ad bellum* and on the *in bello* levels. It declared war—*Jihad*—against all heretics, including many Muslims who hold the wrong beliefs or practices according to its view. On its way to establishing the 'Islamic State', first in Iraq and Syria and then in other areas, ISIS has shown itself willing to carry out the most gruesome atrocities, which need not be described here. By violating its contractarian duties, ISIS loses the right to complain when civilians who contributed to ISIS's goals are targeted. The pre-contractual moral standing of these civilians does not differ from the moral standing of militants who actually do the fighting. The political, religious, and ideological leaders if Isis are legitimate targets for attack no less than its militants and murderers.

It might be thought that the scope of the remedial right to retaliate is wider when the NSA finds shelter among a collective that it assumedly represents; namely, when realizing the right to self-determination of this collective is the NSA's raison d'être. In cases like these, the destruction inflicted upon the collective might convince the relevant organization to cease its violation of the *in bello* rules. Does it follow from our interpretation of the war agreement that such destruction is permissible? Does the targeting of fully uninvolved civilians in order to coerce the organization that fights for them follow the *in bello* rules?

We believe not. Again, the rights of children, the sick, or the elderly may be violated intentionally only in supreme emergencies, a title that seems too strong to describe the threat posed in these cases.[11] Presumably, under less extreme circumstances, decent states should undertake a duty to retaliate by targeting only involved civilians. Deliberate attacks on civilians should be banned by the contract even if, rarely, they succeed in convincing the other side to stop the violations of

---

[10] For the legal constraints on reprisals, see Dinstein, *War, Aggression and Self-Defence* 194–202; Frits Kalshoven and Liesbeth Zegveld, *Constraints on the Waging of War: An Introduction to International Humanitarian Law* (Cambridge: Cambridge University Press, 2011), *passim*; and Andrew D. Mitchell, 'Does One Illegality Merit Another? The Law of Belligerent Reprisals in International Law', *Military Law Review* 170 (2001), 155–77.

[11] Some politicians have argued in the years following 9/11 that the United States is in a condition of supreme emergency, hence in need of completely new (and more permissive) rules to regulate war. For some references and for criticism of such irresponsible use of this concept, see Brian Orend, *The Morality of War* (Peterborough, ON: Broadview Press, 2006), 144–6.

Civilian Immunity. Such attacks are dangerous, because they convey the message that the retaliator wishes to abandon the contract rather than the message that he is willing to re-establish it. They are also dangerous given the likelihood of a misguided determination on the part of the just side that killing uninvolved civilians is necessary.

Just humanitarian intervention might seem as if it is governed by pre-contractual morality. Consider a rogue state that exterminates members of a minority struggling for self-determination. The *jus ad bellum* code, through the 2(4) prohibition, confers on states a right against military intervention by another state. Yet, rescuing this minority from extermination conducted by its own rogue state seems not only morally permissible, but mandatory; hence, humanitarian intervention seems obligatory.[12] True, states usually have a contractual right against military intervention; yet, it might be thought that under the circumstances just described, the war agreement collapses. The rogue state systematically violates the rights that the *in bello* agreement protects, and therefore it simply loses this contractual right.

But this analysis is mistaken: the war agreement never collapses. The rogue state does have a contractual right against intervention of state or non-state actors. The reason for this is the following. The contract still regulates the behaviour of other states because of the high risk of error in establishing the relevant facts about civil wars and in predicting the outcome of military intervention. In such cases as well, the attempt to follow pre-contractual morality (which would allow everyone to go in and rescue the victims) would usually lead to a weaker protection of rights rather than to a stronger one. To be sure, in certain cases, pre-contractually, anybody who can intervene in order to prevent crimes against humanity has a right (and probably an obligation) to do so, namely, to violently cross the borders of a rogue state in order to put an end to a moral disaster. Nevertheless, the temptation to demonize a rival and then fight such a war as a pretext to promoting one's own interests is too strong. Therefore, decent states should agree on a rule that prohibits individual states from entering wars of humanitarian intervention unless they are authorized to do so by a wide coalition of states. Such a coalition would be less likely to make errors and would reduce the risk of countries trying to advance their own interests under the guise of offering humanitarian help. Under the contract, no individual state has a right—and, *a fortiori*, an obligation—to make its own decision to wage a war of humanitarian intervention. Rather, it has a right (and an obligation) to join other states (or the United Nations) in an attempt

---

[12] For the question of whether (just) wars might be optional or are necessarily mandatory, see Noam Zohar, 'Can a War Be Morally "Optional"?' *Journal of Political Philosophy* 4/3 (1996), 229–41, and Kieran Oberman, 'The Myth of the Optional War: Why States Are Required to Wage the Wars They Are Permitted to Wage', *Philosophy and Public Affairs* 43/4 (2015), 255–86.

to establish the relevant facts and, if the coalition of states decides that war is appropriate, to join it.

To conclude, then, although states under conditions of minimally just symmetrical anarchy recognize the possibility of attacks by a rogue state or NSA, they undertake a commitment not to initiate war against it on their own unless the aggression is directed against them. When the aggression is directed against other states or collectives, states undertake a commitment to fight together through a wide coalition or under the aegis of the United Nations.

The analysis offered above strongly suggests that parties to the contract agree to treat their enemy as if it were decent, and that the evidence that they must obtain in order to regard a state or a non-state actor as illegitimate must be decisive. As we read them, the rules that govern retaliations reflect a continued and consistent commitment to rules that govern war against decent stats. The system thus expresses a commitment to treat the violator as decent and the violation committed as an exception.

However, the war agreement leaves room for the possibility that actors are or might become indecent. In the next section, we suggest that states devise a rule whose aim is to deter states from becoming rouge and genocidal, and a rule that enables them to address genocidal aggression. We suggest that the parties address the possibility of indecency by agreeing to make an exception to the otherwise close-to-absolute prohibition against deliberately attacking civilians. The 'balance of terror', created by the mutual conditional threat of indiscriminate slaughter that Western and Communist countries posed to each other, is a manifestation of this norm.

7.3 Supreme Emergencies

Supreme emergencies are cases in which the enemy is ruthless not only in the sense of systematically violating the *jus in bello* rules, but also in terms of its ends—typically the destruction of a political community. According to Walzer, targeting innocent civilians might be morally tolerable in facing such extreme circumstances. We shall refer to this special permission as the supreme emergency exemption, or simply 'the Exemption'.

The nature of the circumstances that trigger the Exemption in Walzer's view is a bit unclear. In introducing the doctrine, he talks about a threat of 'enslavement or extermination'[13] as a condition for defining a situation as a supreme emergency. Such language fits the example he offers of Britain's war against the Germans after the Nazi occupation of Western Europe in 1940. However,

---

[13] Walzer, *Just and Unjust Wars*, 254.

elsewhere, he gives the impression that a supreme emergency occurs any time a country faces defeat that is expected to lead to a loss of sovereignty.[14] In both scenarios, a country is about to be defeated; but in the first scenario, the defeat is expected to culminate in the extermination or enslavement of the defeated nation, while in the second, the loss is expected to be merely political.[15]

We shall focus on the first scenario. Such a situation meets three conditions: (a) some country (or some group) faces a threat of extermination or enslavement; (b) conventional (diplomatic or military) means are unable to counter the threat; and (c) unconventional means, particularly indiscriminate attacks against the enemy, carry a reasonable chance of preventing the otherwise inevitable catastrophe.

Is the threatened country released from the fundamental pre-contractarian constraint against killing the innocent? On the one hand, Walzer seems to capture the prevalent sense that if the heavens are about to fall, there is hardly any limit to what we would do—and would be *entitled* to do—to defend ourselves. Those who declare that they would rather see the extermination of their own people than bring about death and destruction to civilians on the enemy side are often seen as disingenuous or hypocritical. On the other hand, it seems that this is exactly what deontological morality is about: deliberately killing the innocent is impermissible even if it is necessary for bringing about the greater good. Hence, justifying such measures is truly challenging.

Pre-contractual morality has several routes to offer in helping to meet this challenge, but none seems promising. The first is to say that its commitment to human rights is not absolute; namely, that above some threshold it permits—maybe even mandates—a shifting of gears into consequentialist reasoning. In this vein, Walzer argues that when the very existence of a community is at risk, 'the restraint on utilitarian calculation must be lifted'.[16] Later, reflecting on this argument, he speculates that 'perhaps it is only a matter of arithmetic'.[17] Recent critics of Walzer have also interpreted him this way, as if in supreme emergencies he replaces his rights-based outlook with a utilitarian one.[18]

---

[14] Ibid. ('Can soldiers and statesmen override the rights of innocent people *for the sake of their own political community?*...I am inclined to answer this question affirmatively, though not without hesitation and worry' (emphasis added). For an analysis of Walzer's ambivalence on this point, see David C. Hendrickson, 'In Defense of Realism: A Commentary on Just and Unjust Wars', *Ethics and International Affairs* 11 (1997), 19–53.

[15] For a defence of a modified version of the Exemption, see Igor Primoratz, 'The Morality of Terrorism', *Journal of Applied Philosophy* 14 (1997), 221–33, and 'State Terrorism and Counter Terrorism', in Igor Primoratz (ed.), *Terrorism: The Philosophical Issues* (Basingstoke: Palgrave, 2004), 113–27. For a powerful critique, see Cecil A. J. Coady, 'Terrorism, Morality and Supreme Emergency', *Ethics* 114 (2004), 772–89 and Cecil A. J. Coady, 'Terrorism and Innocence', *The Journal of Ethics* 8 (2004), 37–58.

[16] Walzer, *Just and Unjust Wars*, 228.

[17] Ibid., 254.

[18] See Brian Orend, 'Just and Lawful Conduct in War: Reflections on Michael Walzer', *Law and Philosophy* 20/1 (2001), 25; Darrell Cole, 'Death Before Dishonor or Dishonor Before Death? Christian Just War, Terrorism, and Supreme Emergency', *Notre Dame Journal of Law, Ethics and Public Policy* 16

Walzer's reasoning, however, is not really utilitarian. First, according to utilitarianism, individuals or collectives are allowed to give priority to their own lives over the lives of others only if such preference leads to better results *overall*. Yet overall, saturation bombing of the enemy might lead to more death and destruction than would be prevented by such attacks, especially if the aggressor state is much larger than the victim (the state that wants to help itself to the Exemption). From a utilitarian perspective, better that X innocent people be killed than 2X, regardless of the side on which they find themselves in some armed conflict.[19] Yet surely, *if* countries have a right to breach the common rules in order to block a threat of the magnitude of extermination or enslavement, then this right applies even—or *particularly*—to relatively small countries against relatively large ones.

Worse, utilitarianism seems to be insensitive to the national identity of the people whose killing is assumedly necessary to save the attacked country from the expected catastrophic results of defeat. Hence, for example, if the Allies could have won the Second World War by killing thousands of innocent Swedes (who, say, innocently blocked the way to Hitler), that would have been just as permissible as killing thousands of Germans. However, the Exemption seems to have an in-built national partiality.

At this point, one might concede that the Exemption is not based on utilitarian thinking and try instead to base it on the right to self-defence. But this won't do either, because the right to self-defence does not license the killing of innocent bystanders. One might then deny the innocence (in the required sense) of enemy civilians on the basis of McMahan's idea that 'if the Pursuer were *in some measure responsible* for the unjust threat she poses, that would establish an obviously relevant moral asymmetry between you and her and would constitute a sufficient basis for the permissibility of your killing her if that were necessary to defend your life'.[20] Since many enemy civilians are 'in some measure' responsible for the unjust war carried out by their country—definitely more responsible than civilians of the just side that otherwise will be killed—what follows is that the number of people who might qualify as legitimate targets is significantly extended.

This defence of the permission to deliberately target civilians under circumstances of supreme emergency faces two main difficulties. The first, which we cannot elaborate on here, is that the general theory of self-defence that it relies on

---

(2002), 97; Alex J. Bellamy, 'Supreme Emergencies and the Protection of Noncombatants in War', *International Affairs* 80/5 (2004), 838; and Christopher Toner, 'Just War and the Supreme Emergency Exemption', *Philosophical Quarterly* 55 (2005), 549 ('The central argument for the supreme emergency exemption is straightforwardly an argument from consequences').

[19] For the sake of argument, let us assume that the indirect results of blocking the threat (especially in terms of deterring other potential aggressors) do not change the utilitarian calculus in favour of the small country.

[20] McMahan, 'The Ethics of Killing in War', 720 (emphasis added). For simplicity, we replaced 'pursuer' with 'attacker'.

is problematic.²¹ The second is that even if this theory were accepted, it would not cover all targets of killing in supreme emergencies, because many targets, such as young children or opponents of the regime, do not even bear indirect or circumventive responsibility for their country's unjust policy.

Perhaps the self-defence-based explanation of the Exemption could be rescued by relying on the Doctrine of Double Effect and regarding the truly innocent targets as regrettable side-effects of an assumedly justified attack.²² However, it is unclear what the DDE has to say about dropping a bomb on a group of people knowing that some are legitimate targets while others are not. The Doctrine cannot be trusted to save the argument.²³ Indeed, the combination of the minimal responsibility argument and DDE is quite disconcerting, especially when democracies are concerned. It seems to permit saturation bombing of major cities on the grounds that the victims are either supportive of the regime and, therefore, legitimate targets by virtue of being somewhat responsible for the threat, or they are not, in which case their killing is an unintended side-effect of an attack on a legitimate target, and, as such, permissible under DDE.

Pre-contractual morality might try to rely on other arguments to justify the Exemption; but, as shown elsewhere, none is convincing.²⁴ If you stick to your fundamental commitment to human rights, you are committed to standing by while the bad guys wipe out or enslave entire peoples, even if you are a political leader who could rescue your own people by a deliberate killing of innocent unjust civilians.²⁵ If you refuse to stand by, and instead select the only course of action that can prevent the catastrophe, you would be denying all that is morally dear to you.

Contractarianism offers a way out of this impasse, to which we now turn.

7.4 A Contractarian Account of the Exemption

We have seen that a war agreement should devise rules that address a violation of the *in bello* agreement by decent parties. Indeed, these rules are crucial, despite the fact that *ex ante*, decent states prefer an outcome in which the agreement is observed to an outcome in which their war is governed by pre-contractual

---

²¹ See Benbaji, 'The Responsibility of Soldiers and the Ethics of Killing in War' and Seth Lazar, 'Responsibility, Risk, and Killing in Self-Defense', *Ethics* 119/4 (2009), 699–728.

²² See McMahan, 'Who is Morally Liable to be Killed in War' and Lazar, 'Liability and the Ethics of War'. The idea of using DDE to respond to supreme emergencies is developed by Bellamy, 'Supreme Emergencies and the Protection of Noncombatants in War', and is explicitly rejected by Rawls, *The Law of Peoples*, 104–5.

²³ For a discussion of these issues, see McMahan, 'Self-Defense and the Problem of the Innocent Attacker', 252–90.

²⁴ See Daniel Statman, 'Supreme Emergencies Revisited' *Ethics* 117 (2006), 58–79.

²⁵ Christopher Toner explicitly prefers to this option in 'Just War and the Supreme Emergency Exemption'. Toner argues that 'we and the world will perish anyway, sooner or later; we cannot change that. But we can at least, in so far as in our power, let justice be done while we live' (561).

morality. The contracting parties should take another possibility into account. Namely, the possibility that decent states may become indecent and strive for ends that are manifestly immoral, such as ethnic cleansing or genocide. To prevent such grave moral crimes, the contracting parties reach an agreement that confers remedial rights, whose aim is to deter the parties from engaging in such projects in the first place. They agree that when collectives face a threat of genocide against which they cannot protect themselves by conventional measures, they are released from the otherwise close-to-absolute prohibition against deliberately attacking civilians. They agree on rules that license the use of WMD (as a last resort only!) against the perpetrators of such evils.

This licence is supposed to have two desired effects: first, to deter states from grave moral crimes like mass killing of innocents; and second, to provide citizens of potentially rogue states with a powerful motivation not to elect leaders who have such immoral plans in mind, and to make a serious effort to struggle against them if they seize political power (via democratic elections or via use of force). Indeed, a regime that contemplates carrying out actions such as mass murder typically oppresses its own citizens, which means that opposing it carries significant risk. Precisely for this reason, its citizens must be provided with an especially powerful counter-motive that could balance the threat imposed by their own leaders. They must realize that if the genocidal plans of their government materialized, the price that they—the citizens and their families—would be forced to pay would be terrible. In other words, citizens are more likely to struggle for the decency of their state in a world in which a supreme emergency exemption is a universally accepted norm. Under a threat of unrestrained war—a war released of almost all moral restrictions—the risk of enslavement and ethnic cleansing is reduced.

The idea that intentional killing of the truly innocent might be (contractually) permissible may still seem outrageous. To make it easier to accept, recall that according to contractarianism, a similar right to kill the innocent has already been acknowledged; namely, the (contractual) right to kill non-culpable, non-responsible combatants. In supreme emergencies, the category of legitimate targets is widened to include not only innocent combatants but the young, the sick, the old, and so on. Note further that revisionist morality allows the conscious killing of the innocent, if the killing is proportionate. Most revisionists further concede that when the numbers are sufficiently high, deliberately killing innocents might be permissible: while using the fat man and killing him in order to save five people might be impermissible, doing so in order to save 100 people might well be permissible.

No parallel reason exists to expose ourselves to total wars in order to increase the incentive of citizens to prevent governments from launching regular (i.e. non-genocidal) unjust wars. Unlike enslavement and ethnic cleansing, whose evil is easily recognized, regular wars are marked by reasonable disagreement over the justness of their cause. Therefore, requiring citizens to block crimes against peace would be ineffectual.

Under our interpretation, the contract devises three categories. When one side violates merely conventional rules, the other side is no longer subject to these rules. In terms of its deliberative duties, the retaliator is not obliged to make sure that the retaliatory action is effective either in advancing victory or in pushing the other side to re-enter the contract. The retaliating side may rely on the general presumption that retaliations have desirable effects. The second category is the *in bello* rules that are tied to pre-contractual morality in a clearer way, like the rules that prohibit targeting civilians. Under regular circumstances, targeting uninvolved civilians is impermissible even if it can reasonably be expected to be effective in advancing victory or in renewing the violator's commitment to the contract. The supreme emergency exemption forms a third category. Exceptionally, it allows the targeting of innocents. However, if counter-violation under extreme circumstance cannot be expected to be effective, *it is morally out of the question*.[26] Intentionally killing the innocent just for the sake of revenge is neither allowed by pre-contractual morality nor licensed by the war agreement.

The basic structure of the contractarian argument for the Exemption is as follows. Citizens authorize their states to act on their behalf (or on behalf of their children) in the international realm; a war agreement that includes the Exemption is in their interest; therefore, states accept the agreement on behalf of their subjects. By accepting the agreement, citizens lose the pre-contractual right not to be targeted in case their country poses a genocidal threat to some other group, which cannot be countered by regular, conventional measures. The introduction of the supreme emergency exemption relies on two assumptions: (a) that if states know that moral crimes like genocide on their part might lead to counter-violations by their enemy, this might deter them from committing these crimes; and (b) that if they have already carried out such violations, they will consider refraining from further acts if faced with similar violations directed against them. A balance of terror, so to speak, may be mutually beneficial.

Walzer's reasoning seems to reach the same conclusion. The danger of radical evil makes a world in which decent states hold WMD safer than one in which they do not:

> It is not tolerable that advances in technology should put our nation, or any nation, at the mercy of a great power willing to menace the world or to press its authority outwards in the shadow of an implicit threat...Against an enemy actually willing to use the bomb, self-defense is impossible, and it makes sense to say that the only compensating step is the (immoral) threat to respond in kind.

---

[26] In Walzer's words, what states are allowed to do in supreme emergencies is 'whatever is militarily necessary to avoid the disaster'. See *Arguing About War*, 40.

7.4 A CONTRACTARIAN ACCOUNT OF THE EXEMPTION 177

Hence any state confronted by a nuclear adversary... and capable of developing its own bomb is likely to do so, seeking safety in a balance of terror. Mutual disarmament would clearly be a preferable alternative but it is available only to the two countries working closely together, whereas deterrence is the likely choice of either one of them alone.[27]

In this passage, Walzer describes a simple dilemma generated by problems of trust. While all parties prefer mutual nuclear disarmament to a balance of terror, each prefers a situation in which it alone possesses WMD, while the others have no such weapons. Since there is no common force that could assure that states actually disarm from WMD, no party can trust the others to do so. Hence, the most rational course of action would be for each state either to acquire WMD or to secure the protection of an ally who possesses them. Morally restrained, rational, and self-interested states living in a world that is burdened by the availability of nuclear technologies would prefer to be protected by WMD to living without such protection. Alas, parties that manufacture or buy WMD might become murderous. Arguably, the safest way to deter them from becoming genocidal (by using such weapons) is by making clear to them that we have WMD too and are willing to use them in case they do. It is hard to imagine a way out of this equilibrium.

We might think, though, that the right to possess WMD and the remedial right to strike back do not satisfy *Fairness*. The contract could be said to discriminate against weak states with no WMD capabilities because they do not benefit from the permission to use the kind of weapons that is granted in extreme circumstances. This concern is ungrounded. The Exemption is fair, first of all, as it is one of the measures by which a fair arrangement is enforced. It is an essential part of an arrangement that prohibits the use of force and secures states' ability to enforce the prohibition with minimum casualties to innocents in a world in which states are free to hold arms and might become evil. The Exemption is also fair in the more standard sense articulated earlier; namely, that it cannot be known in advance to generate inequality between the weak and the powerful. Even weak states can obtain or manufacture dirty bombs that can pose a serious threat to their enemies. Moreover, weak states can and often do form alliances with strong ones, which takes the sting out of the unfairness objection.

We turn now to *Actuality*. Has the international community actually accepted the proviso that, in supreme emergencies, states are allowed to breach the rules restraining the use of force in non-emergency situations? Although states have not explicitly accepted an exemption that allows direct targeting of the innocent, they seem to have tacitly accepted it. The approach to nuclear deterrence, especially

---

[27] Walzer, *Just and Unjust Wars*, 273–4. This position is controversial. See, e.g., Douglas Lackey, 'The Intentions of Deterrence', in Steven P. Lee and Avner Cohen (eds.), *Nuclear Weapons and the Future of Humanity: The Fundamental Questions* (Totowa, NJ: Rowman and Allanheld, 1984), 307–19.

during the Cold War period, is a clear manifestation of this acceptance. In those years, a 'balance of terror' was produced by the mutual conditional threat of indiscriminate slaughter that the Western and Communist countries posed to each other. Western governments threatened to kill fully innocent people in Eastern Europe in case their own citizens were attacked with nuclear weapons by Communist countries, and the other way around.[28] Thus, we claim that by tolerating and imposing the threats that maintain the balance of terror, the society of states effectively accepts the Exemption.

The contractarian account of supreme emergencies explains the moral distinction mentioned above between innocent citizens of an evil state (citizens of Nazi Germany, for example), who may be targeted under such circumstances, and citizens of neutral states (Swedish citizens, for instance), who may not. Both considerations mentioned above in favour of the Exemption apply only to evil states. Neutral states did not violate the contract; therefore, no rule allows their destruction. To deter evil leaders from committing large-scale crimes in war, it won't help to threaten to kill innocent citizens of *other* countries. Similarly, to give citizens of evil states an incentive to oppose their regime, it won't help to license the indiscriminate killing of citizens residing in neutral countries.

Moreover, the Exemption applies to all evil states, no matter how big they are. The rule that allows striking back is accepted in order to motivate all political communities to radically oppose evil leadership. Hence, small states are allowed to defend themselves from genocide even if the defensive action involves killing more innocents that would otherwise be killed by the rogue (and bigger) states.

7.5 Conclusion

According to the theory we put forward in this book, the ethics of war is shaped by an agreement between decent states (and NSAs) regarding the circumstances under which they may resort to war and the rules that apply to them once wars break out. One of the basic terms of the war agreement confers on states an inherent right to self-defence against violators of the *jus ad bellum* prohibition on the use of force. The question around which this chapter revolved was how decent actors should fight against parties that do not subject themselves to the *in bello* agreement.

---

[28] See 'Legality of the Threat Or Use of Nuclear Weapons', Advisory Opinion of 8 July 1996, where the ICJ states that: 'The emergence, as lex lata, of a customary rule specifically prohibiting the use of nuclear weapons as such is hampered by the continuing tensions between the nascent opinio juris, on the one hand, and the still strong adherence to the doctrine of deterrence (in which the right to use those weapons in the exercise of the right to self-defense against an armed attack threatening the vital security interests of the State is reserved), on the other' (http://www.icj-cij.org/docket/files/95/7497.pdf), 97.

The natural assumption is that when the war agreement is violated systematically by one of the parties, it loses its binding force and the parties revert back to the state of nature. For Hobbesians, that would mean a state of affairs in which there are no binding moral obligations. For Lockeans like us, it would mean a state of affairs in which the parties are subject to pre-contractual rights and duties, often referred to as deep morality. We have argued, however, that this assumption must be resisted. The normative implications of breaching the war contract are themselves part of the contract. In the absence of an international body that could enforce the contract on those violating it, the parties must have the right to respond to violations of the *in bello* rules. These remedies—these counter-violations—should be cautious and restrained, with a constant eye on the goal of reinstituting the parties' commitment to the broken rules.

We offered three categories. The permission granted by the contract to break the *in bello* rules that the enemy systematically violates applies mainly to prohibitions whose source is merely conventional, such as those outlawing certain weapons. Second, retaliation might involve breaking the rules that prohibit attacks on civilians responsible for the enemy war effort, such as politicians, officials, or party members. This permission never extends to the paradigmatically innocent—the young, the sick, the old, and so on—except in cases of supreme emergency, which give rise to the Exemption.[29]

We might expect that the second category would apply in particular force to organizations like ISIS, whose goals and methods are morally intolerable. However, with such organizations, it is often the case that no civilian targets banned by contractual morality comprise a target whose attack would be effective in defeating them. In other words, 'taking off one's gloves' in fighting against ISIS would probably be ineffective, hence would be usually ruled out. Things are different with organizations that act from within some ethnic or religious group and who see themselves as representing the group and as acting on its behalf. In these cases, attacks on contributing civilians or on civilian infrastructure might be effective and hence licensed by the contract.

Finally, a standard complaint is voiced in the last decade against the current legal and moral rules regarding war which, it is argued, have left decent states with no effective way of defending themselves against rogue states or NSAs. Presumably, these rules are outdated and in need of serious revision.[30] The concern underlying this complaint is genuine; the rules governing war should be interpreted as terms in an agreement whose aim is to enable the parties to develop an effective way of defending themselves against perceived violations of the

---

[29] Or what has been called elsewhere, 'special permissions'. See Statman, 'Supreme Emergencies Revisited'.

[30] See Michael Walzer 'Coda: Can the Good Guys Win?', *The European Journal of International Law* 24/1 (2013), 433–44, responding to reports of soldier complaints about the rules of engagement in Afghanistan.

prohibition on the use of force. Hence, if indeed the morality of war left parties with no such mechanisms, this would constitute a serious objection to it. Fortunately, this is not the case. When taking off one's gloves can reasonably be hoped to aid in the defeat of one's enemy, doing so is permissible in response to enemy violations of the *in bello* code (under the constraints explicated above). When it cannot, then doing so remains impermissible, but that does not justify any special complaint.

As we saw in Sections 3.5, 6.7 and in this chapter, the 'new wars' phenomena invite four main questions. First, how does the *ad bellum* agreement regulate disputes in which one of the parties is a stateless nation that demands political independence in a territory to which it has no positive legal right? Second, how does the *in bello* agreement regulate wars where one of the parties did not or could not institute a professional army, and, consequently, did not establish a distinction between soldiers and civilians? Third, how does the *ad bellum* contract regulate conflicts in which one of the parties is indecent, namely, has little respect for pre-contractual morality? Finally, how does the *in bello* agreement apply to a conflict in which one of the parties systematically violates the *in bello* agreement?

Shortly after the publication of Walzer's *Just and Unjust Wars*, critics argued that the changing character of war renders Walzer's 'statism' irrelevant because his traditional approach cannot address any of these questions.[31] More recently, in a *New York Times* Op-Ed, McMahan argued that one reason for the collapse of the traditional just war theory is its statism: 'Most recent wars have not been of the sort to which [Walzer's] theory most readily applies—namely, wars between regular armies deployed by states. Many have instead been between the regular army of a state and "rogue" forces not under the control of any state.'[32] These types of conflict, McMahan contends, 'are resistant to moral evaluation within the state-centric framework of the traditional theory'. Revisionist 'individualization'—the attempt to apply the morality of killing in self- or other-defence in domestic society to the morality of war—is partly a response to the new wars phenomenon.[33]

Together with Section 3.5—in which we advanced a contractarian *ad bellum* of wars of independence—and Section 6.7—where we outlined a contractarian *in bello* code for asymmetric wars—this chapter aimed to show that the 'individualization of war' is too hasty a response to the phenomenon of irregular wars.

---

[31] Most notably Luban, 'Just War and Human Rights'.
[32] Jeff McMahan, 'Rethinking the "Just War"', Part 1, *New York Times Opinionater*, November 11 2012. (https://opinionator.blogs.nytimes.com/2012/11/11/rethinking-the-just-war-part-1/).
[33] A term coined by Gabriella Blum, 'The Individualization of War: From War to Policing in the Regulation of Armed Conflicts', in Austin Sarat, Lawrence Douglas, and Martha Merrill Umphrey (eds.), *Law and War* (Stanford, CA: Stanford University Press, 2013).

8
Concluding Remarks

8.1 Overview of the Argument

(I) The idea of universal human rights expresses the recognition that all human beings have intrinsic value. Their moral standing poses constraints on how they ought to be treated. In particular, they ought not to be treated as mere means to the achievement of contingent goals, but rather, in Kant's famous words, as ends in themselves. However, human beings can waive some of these rights, under well-defined circumstances, in exchange for various benefits. Using the guiding myth of the social contract tradition, we can say that by organizing themselves into a society, rather than existing as a bunch of atomized individuals, human beings create a new division of moral rights that better serves their interests and the overall fulfilment of their pre-contractual rights. Such a contractual distribution of rights and duties takes place only if the social arrangement it creates does not solidify or bring into being unjust relationships; in short, only if it is fair. Social arrangements that meet these conditions are morally effective—they determine a conventional distribution of moral rights and duties—once they are actually accepted within the relevant community.

Thus, people have the moral power to waive natural or pre-contractual rights in exchange for various benefits and a better protection of rights. More accurately, at least where fundamental rights are concerned, they don't waive them vis-à-vis *all* people, but vis-à-vis specific role-holders, under well-defined circumstances. For instance, by habitually following the social rules by which a society is organized, society members lose the right they have vis-à-vis policemen not to be arrested before the policemen verify first that the arrest order is justified. Similarly, prisoners lose the right they hold vis-à-vis the prison guards not to be locked up in their cells before the guards verify that the sentence was warranted, and so on. Waiver of rights in these contexts is performed by transferring the responsibility for respecting the rights in question (i.e. the right of the innocents not to be locked up) from the role-holder implementing the relevant state decision (the prison guard) to the one making it (the judge). Such a waiver does not absolve all social institutions from their responsibility to respect the rights in question.

The feature of morality to which we are drawing attention is not at all marginal. Human beings hardly ever interact outside organized society, in an imaginary void beyond (or beneath) the reach of social conventions and institutions. And living within a well-ordered society involves accepting the arrangements that govern it.

Waiver of rights is a permanent feature of social life. Asking what people would owe each other if they existed outside any social or political framework has little value in attempting to discover the duties to which they are in fact subject and the rights that they possess. The normative world in which humans live is shaped by social conventions and institutions that define not only the social and legal rules to which they are subject, but also the moral duties they bear and the moral rights they hold.

Two reservations regarding the moral framework employed in these chapters are in order. First, we had no intention of providing a full justification of moral contractarianism—a task that would take us far beyond the boundaries of the present book. In Chapter 2, we merely provided some of its basic rationale and, in subsequent chapters, we showed its fruitfulness in thinking about the morality of war. Second, we did not attempt to address all the difficulties that our approach raises: just as writers who propose a utilitarian defence of the ethics of war cannot be expected to deal with all the theoretical challenges utilitarianism faces, so writers relying on contractarianism cannot be expected to present a comprehensive theory of the latter either.

(II) Roughly, the morality of war advanced in this book can be divided into three main themes. First, the society of states under conditions of minimally just symmetrical anarchy should agree to the Charter's Article 2(4) prohibition against first use of force by an individual state against the territorial integrity of another state or against a stateless nation in a non-violent struggle for political independence. Under the *ad bellum* contract, states should use other means to pursue pre-contractual justice, or to realize their pre-contractual right to fair distribution of territories and resources. The 2(4) prohibition is enforced by self-help through the Article 51's right to self-defence. States are allowed to use force against other states that violate the contractual right against aggression. In the absence of a common power that could enforce the *ad bellum* agreement, states should allow each other to use force against violators of the agreement. The contract confers on states a defensive right to protect their contractual right to freedom from aggression by self-help, and to realize this right by use of force. The scope of the right to use force in self-defence is quite narrow. In order to make sure that defensive rights are not abused, the law permits war against threats that are both substantial and visible; that is, imminent threats to the territorial integrity of a legitimate state.

Thus, on the one hand, states are barred from waging some wars that would be pre-contractually permissible, like preventive or subsistence wars or wars of humanitarian interventions. On the other hand, they are allowed to wage some wars that would be pre-contractually forbidden, like national-defence wars aiming at the protection of territorial rights. Moreover, the contract permits (pre-contractually) disproportionate war, in which the expected moral price of surrender is lower than the expected price of going to a permissible defensive war. Both the prohibition to use force in order to resolve conflicts and the right to wage war in

8.1 OVERVIEW OF THE ARGUMENT 183

response to forbidden use of force by an aggressive state are valid independently of facts about the ultimate truth of the relevant claims made by the parties. The *ad bellum* agreement outlaws any use of force whose aim is conflict resolution, even for a state that has a justified grievance against its adversary and can enforce its rights by force only. Likewise, the right to use military force in order to defend borders does not assume that the borders protect any moral value (like self-determination, way of life, etc.).

Second, states are entitled to have obedient armies as means to address aggression. This entitlement empowers states to institute the role of soldiery through the legal duty of obedience. The agreement does so by subjecting Just and Unjust combatants to the same *in bello* rules. Thanks to their legal equality, soldiers are legally entitled to participate in a war fought by their states, without making sure that the war is just.

Furthermore, to enable states to defend themselves, the *in bello* contract contains a permissive *in bello* necessity rule: it confers on armies a right to destroy enemy military forces by targeting all enemy soldiers and all enemy military objects. Armies are entitled to presume that eliminating *any* military object promotes military victory—that is, submission of the military forces of the enemy—and deters it from future violation of the 2(4) prohibition.

The third element follows from the first two. If the agreement is accepted, armies are the tools by which states defend themselves against unjust aggression. Assuming that the agreement is accepted, states can, and therefore should, agree to grant immunity from intentional attack to both involved and uninvolved civilians, and allow only proportionate collateral harm to them. This is because, presumably, *Civilian Immunity* would not change their chances of victory, and would make the military clash less bloody and more respectful of human rights.

We argued that the fact that this arrangement is accepted within the society of states entails that if this arrangement is fair, it is morally effective. That is, if the war agreement can be presented as a fair contract between decently partial states existing within a minimally just symmetrical anarchy, it determines the distribution of the fundamental (moral) rights of civilians and soldiers.

As we repeatedly stressed, under our interpretation, the war agreement is accepted on the assumption that states exist under conditions of minimally just symmetrical anarchy. A society of states is anarchical if there is no common authority to adjudicate conflicts between them (and between states and non-state actors). The anarchy is symmetrical if, for most future conflicts, states can know in advance that they will do better by negotiating with their adversaries rather than fighting them. The anarchical society of states is minimally just if, for reasons of normative uncertainty, most wars whose aim is the implementation of pre-contractual justice are impermissible; had states been fully impartial, they would have avoided using force in resolving their conflicts.

Is the war agreement (signed by parties existing within a minimally just symmetrical anarchy) mutually beneficial and fair? Consider first the *ad bellum* contract. We argued that under such circumstances, states have a strong reason to waive their right to wage wars whose aim is the implementation of pre-contractual justice. They should waive this right despite the fact that in rare cases they can be certain that the use of force is necessary in order to enforce their pre-contractual rights, and that the harm caused by the war will be a lesser evil. They should enter a contract prohibiting the use of force as a way to resolve conflicts in *all cases*.

Is such an *ad bellum* arrangement fair? A blanket prohibition on the use of force requires poor states to refrain from using force against rich states in order to obtain resources that are necessary for a minimally dignified life or even for mere survival. War is impermissible even if the inequality between the poor and rich countries is a result of past injustice. Furthermore, the agreement leaves oppressed groups in rogue states with a weaker protection than that to which they would be entitled according to pre-contractual morality. Finally, stateless nations struggling for independence are not allowed to use force (unless they are attacked). The *ad bellum* contract, therefore, seems to solidify unjust relations, which are the essence of unfairness as defined earlier. The contract is nevertheless fair in one clear sense. The results of wars, especially those fought for the sake of redistributing resources or preventing intra-state oppression, are radically uncertain. Hence, under minimally just background circumstances, mistakes about what pre-contractual justice requires in the international realm are very likely. From a moral perspective, it is almost always better to avoid wars. In light of this fact, states and non-state actors (NSAs) should subject themselves to a rule that outlaws any use of military force. This rule ultimately promotes justice rather than compromising it, especially because states are partial and are likely to believe that their pre-contractual right should be enforced by war.

(III) Before moving to the contractarian virtues of the *in bello* contract, let us highlight further important aspects of our analysis of the contractual right to self-defence. We suggest that the basic logic of the *ad bellum* agreement applies to NSAs as well—in particular, to collectives struggling for national independence. As we interpreted it, the contract assumes that for them, just like for states, in most cases, negotiation and compromise are expected to yield better overall results (in terms of welfare and justice) than resort to violence would. Hence, the prohibition against first use of force applies to stateless nations as well. The difference—at the *ad bellum* level—between states and NSAs has to do with the condition that legitimates the use of force in self-defence. While with states this condition is a threat to territorial integrity, with NSAs the threat is violent oppression directed against the group in an attempt to crush its claim to self-determination.

Another important consequence of our analysis is noteworthy. The right to self-defence is the right to use force in order to undo a violation of the territorial integrity or political independence of a just state. But just as importantly, the *ad bellum* arrangement also aims at deterring potential aggressors from violating the prohibition on first use of force. In particular, just states have a right to continue the war against an actual aggressor so as to deter it from further violating the *ad bellum* contract in the near future. It is permissible to impose further costs on the aggressor, to restore its commitment to the contract.

We further observed that an uncompromising war is rarely necessary in order to achieve deterrence. The Clausewitzian idea of 'annihilation' of the enemy military forces may seem attractive, in cases in which this is necessary to prevent future aggression. But the parties to the imagined negotiation realize that they had better resist this attraction. Although in some cases an uncompromising war would yield better and fairer results overall, the war agreement assumes that in most cases it wouldn't. Therefore, decent and partial states have an interest in reaching an agreement that would rule out such wars in *all* cases. Under this interpretation, undertaking the contractual prohibition on use of force implies that the parties undertake a duty to end wars soon after weakening the aggressor in a way that makes further aggression in the near future unlikely. The war should end even if the enemy still possesses military capabilities. The argument against uncompromising wars follows from the *jus ad bellum* prohibition on preventive wars. By committing themselves to a rule that allows only defensive force, the parties reduce the chances of mistakes leading to unnecessary use of force. Another advantage they gain by accepting this term is that they defend themselves from the grim prospect of being the object of preventive or uncompromising wars fought against them.

(IV) We now turn to the arguments for the other major themes in the agreement. Under the war agreement, soldiers are legally equal. Their equality consists of (a) the right of Unjust Combatants to kill Just Combatants, and the right of Just Combatant to kill almost all Unjust Combatants; (b) the right to carry out attacks without verifying their causal contribution to the achievement of victory; and (c) the right to knowingly (though unintentionally) bring about the death of clearly innocent people (enemy civilians) as a side-effect of attacks on military targets.

As individualists convincingly argue, the morality of defensive rights falls short of explaining this feature of the contract. Yet, the legal equality of combatants meets the conditions of *Mutual Benefit* and *Fairness*; hence, it is presumably true that decent states and the individuals they represent freely accept it. As we have just observed, under symmetrical anarchy, no common power exists to enforce the prohibition against the first use of force. States therefore agree on a self-help-based regime that permits wars whose aim is to counter illegal threats. One further (crucial) function of this right is to deter decent states from violating the contractual right against the use of force. Now, the war agreement assumes that just wars and deterrence are more effective if states know that aggression immediately activates the armies of the attacked states in defence of their contractual rights. Hence, obedient armies, whose soldiers are entitled to fight upon command without first needing to be convinced that their war is just, better secure deterrence and effective self-defence. States, therefore, grant each other the right to maintain obedient armies. This arrangement is mutually beneficial; so presumably, combatants freely accept it.

Similarly, since attacks on combatants and on military targets usually contribute to the weakening of enemy military forces, the parties subject themselves to a rule that permits these attacks. Soldiers are entitled to carry them out without making sure that they are indeed necessary for victory. Following a rule that restricts attacks to enemy soldiers who significantly contributed to the perceived unjust attack is costly. Such a rule compromises the ability of the just side to defend itself.

The legal equality of soldiers and their non-immunity is fair, since it cannot be known in advance that it will discriminate against any party. Moreover, it is fair in another sense: it constitutes the most efficient way to maintain a fair agreement.

The war agreement is morally effective if it is accepted by states that institute the role of soldiery, and by soldiers who occupy the role that the society of states designed. The moral equality of combatants—their equal moral right to maim and kill each other—is, therefore, a corollary of the right to hold obedient armies that partial and decent states grant to each other.

The *moral* right of soldiers to engage in morally risky behaviour such as war sounds far-reaching, but basically it is no different from that of other role-holders in society, such as policemen, prison guards, and others, whose jobs involve the risk of violating serious rights. In all these cases, there is good reason to exempt the role-holders from responsibility for the (obviously unintentional) violation of rights, and to transfer the responsibility for such violations onto the political bodies under which they function. The moral responsibility for unjust laws rests on the shoulders of the legislator, not on those of the prison guard or the executioner. Similarly, politicians who initiated a war and (to a lesser degree) the political society that allowed them to do so are morally responsible for it, but not the soldiers who followed their orders.

This normative structure reflects an important feature of our normative powers. When it comes to fundamental human rights, their bearers give them up vis-à-vis specific people under specific circumstances, by transferring the responsibility to respect and protect these rights to other role-holders. The moral division of labour usually meets what we have called the 'conservation of rights violations' condition.

(V) The third element of the war agreement is the blanket immunity conferred on civilians. While states have an interest in granting each other the right to wage war in self-defence, they also have an obvious interest in reducing the horrors involved in war. This dual goal—guaranteeing effective self-defence while reducing the horrors of war—underlies the distinction between combatants and non-combatants. If the *jus ad bellum* contract is accepted, states attack one another and defend themselves from aggression by using military force. Hence, under this regime, (directly) attacking civilians usually does not significantly contribute to victory. To win, the just side must destroy the army of the enemy, at least in part. The agreement further assumes that if armies were allowed to attack politicians, party members, officials, etc. (viz., involved civilians) in the rare cases that doing so might promote victory, this permission would often be misused by error or abused with malice. The parties should therefore agree to rule out this option altogether and undertake an almost unconditional ban on direct attacks against civilians and against civilian infrastructure.

Still, effective fighting depends on relaxing the duty to refrain from harming civilians. Granting immunity to civilians from collateral damage would seriously impede the ability of armies to weaken enemy military forces, especially in wars that involve aerial warfare and in wars fought in proximity to residential areas. The permissions involved in the rule of *in bello* proportionality are necessary to enforce the prohibition on the use of force. Presumably, therefore, the parties to the war agreement freely accept them. Civilians have no right not to be collaterally harmed by combatants of the enemy side, because in habitually following the social rules by which their polity organized itself, they authorize their countries to waive this right on their behalf. In return, their own country obtains the right to carry out military operations even when proportionate harm to civilians is foreseeable.

So far for the main arguments in defence of the three pillars of the agreement that regulates conflicts between decent states: the prohibition on use of force, the permission to harm combatants, and the duty to spare civilians.

(VI) As we read it, the war agreement entails an *ad bellum* code for wars of independence (and for wars which states should fight *as if* they are wars of independence) as well as an *in bello* code for asymmetric wars.

Like Walzer and his followers, we assume that individuals have a pre-contractual right to live in a state whose citizenry has willingly established political cooperation through the political institutions and through the legal system that this state maintains. Alongside the classical pre-contractual rights to life, to bodily integrity, etc., individuals have a pre-contractual right to membership in a political society that maintains their way of life and their national/cultural identity. In many cases, members of a cultural or a national group are governed by political institutions that fail to do so, and in some such cases, they have a right to establish a state in a territory to which they have no recognized legal claim. In rare cases, stateless nations are entitled to use proportionate force to realize this right.

It seems, then, that not only states but also (decent) stateless nations are parties to the war agreement. We suggested that this agreement between states and stateless nations prohibits the use of force, and commands non-violent struggles for independence. Stateless nations have a right to use force only when their non-violent struggle for independence is violently suppressed, in which case a war of independence might be just. The just aim of such wars is not a defence of a contractual right to territorial integrity, a right that stateless peoples don't yet have, but to gain a right to territorial integrity; to establish a state of their own.

How ought wars of independence to be fought? The *in bello* rule that grants immunity from direct attack to the civil society is usually unfair to freedom fighters. For instance, it would be unfair to require them to wear uniform and to be organized in isolated military bases. The power relations are such that the strong side typically will be easily able to target individuals who identify themselves as militants by wearing uniforms. The prohibition on targeting civil infrastructure is also unfair. Freedom fighters cannot significantly weaken the army of a state, and their only chance of achieving the just aim of their war is by influencing public opinion. Contractarianism thus implies that while the *in bello* agreement that governs (asymmetric) wars of independence prohibits targeting civilians, it allows the spread of terror by targeting civil infrastructure.

(VII) The agreement has a fourth dimension. Parties to the contract should take into account the possibility that states and NSAs may be or may become indecent. The parties to the agreement might lose respect for the basic rights of individuals and for the contractual and pre-contractual rights of other states and stateless nations. The contract should devise tools to address the threats posed by parties that have no respect for either pre-contractual morality or the accepted international agreements regulating war. The contract must include special permissions enabling decent states to deter such players and fight them, as well as special permissions allowing them to deter decent players from becoming indecent.

The main mechanism to deter would-be wrongdoers on the *ad bellum* level is the right to state self-defence. States are allowed to go to war to eliminate threats to their sovereignty or to their territorial integrity. In case of NSAs, they have a right to go to war against those who threaten to crush their national or religious rights by force. The normative picture on the *in bello* level is more complex. The agreement devises rules that allow decent parties to deter states and NSAs from indiscriminate killing. Given the very high moral stakes and the doubtful odds that response in kind would be effective, the agreement should not allow indiscriminate killing in response to indiscriminate killing. The possibility of error and escalation are crucial in explaining why this option is ruled out. True, intentional killing might be pre-contractually permissible where its morally important positive effects are especially significant. However, if such permission were granted, it would probably be misused because of misperceptions about what the enemy did and about what exactly the normative ramifications are. Hence, the parties to the negotiation should undertake a rule that almost unconditionally outlaws such attacks, regardless of what the enemy does. That said, they do allow the breaking of *Civilian Immunity*. In retaliating against indiscriminate killing, the parties are allowed to target involved civilians and civilian infrastructure.

Indeed, the normative picture might change in cases of supreme emergency. The contracting parties should find ways to prevent threats of genocide, or the indiscriminate use of WMD, by deterring indecent parties from producing such threats. The potential victims are allowed to threaten to fight back in the same way (and are allowed to arm themselves accordingly). Under extreme circumstances, they have a special permission to kill citizens of the indecent state with WMD indiscriminately. Under radical conditions—where a solid basis exists to assume that actually executing these threats would deter the indecent party and obstruct its evil plan—they may do so. The powerful prohibition against the mass killing of civilians cannot be breached in any other case.

Breaches of the war contract never push the parties back to the state of nature, under the umbrella, as it were, of pre-contractual morality. Rather, the parties to the contract agree not only on the rules regulating war, but also on the rules regulating the responses to violations of these rules. They aim to reduce the risk that decent or indecent parties will violate their contractual rights. This is a crucial theoretical point with important practical implications. The parties should restrain themselves in the face of *in bello* violations and encourage a constant effort to get the other parties to play by the rules. The prudential and moral importance of this recommendation cannot be exaggerated, especially in times of strong pressure from both the public and politicians to 'do something' in the face of patently barbaric attacks. Contractarianism does not divert from pre-contractual morality in its refusal to license widespread attacks on the innocents, especially if these attacks serve no purpose other than revenge.

(VIII) To be morally effective, the social rules by which states ought to regulate warfare should meet *Fairness* and *Mutual Benefit*. If the rules meet these conditions, it is presumably true that individuals who habitually obey these rules freely accept them. By habitually obeying social rules, individuals activate their moral powers to undertake certain duties toward others and to release others from certain duties toward them. In particular, we argued that by accepting their roles, soldiers release other soldiers from the duty not to kill them in war. We further argued that in habitually following the rules by which a political society organizes itself, civilians authorize it to waive some of their rights; for example, the right not to be collaterally harmed in war.

We drew a crucial distinction between two kinds of (fair and mutually beneficial) rules that constitute the agreement. Some rules 'serve' their addressees and therefore have legitimate ('Razian') authority over them. Some rules are mutually beneficial and fair despite the fact that they do not serve their addressees. We summarize this distinction here and reinforce it. To simplify, we shall say that a rule asserting that x must not v has Razian authority over x only if, typically, x must not v *independently of the rule*. Similarly, a rule that asserts that x may v has Razian authority over x only if typically, x may v, *independently of the rule*.

Consider the *ad bellum* rule that prohibits the use of force. The argument presented in Chapter 3 implies that typically, it has legitimate authority over states in minimally just circumstances. This is because generally, if the rule asserts that State S should not use force, S should not do so, independently of the rule. Legislating the rule is important because if the rule is accepted within the society of states, it is more probable that its addressees will avoid impermissible wars; that is, would do what they should do independently of the rule. The same is true of *Civilian Immunity*. The argument presented in Chapter 5 implies that targeting civilians is typically impermissible, whether or not soldiers are subject to the positive law that prohibits it. The fact that the rule is accepted within the society of states makes it more probable that combatants will do what they must do anyway.

But social rules may be mutually beneficial and fair, and thereby morally effective, even if they have no legitimate authority. Indeed, our argument in favour of the rule that permits defensive wars does not imply that this rule has Razian authority. Revisionists insist that defensive wars are often disproportionate. A defensive war might involve killing, while the alternative, surrender, involves merely giving up on certain territory. We do not deny this. Rather, we argued that the rule that permits state self-defence is fair and mutually beneficial because it helps states to enforce a fair and mutually beneficial contract that outlaws use of force. We concede that without the rule that permits state self-defence, this defensive war would have been impermissible. We inferred that defensive wars are permissible *because* presumably, the rule that permits them has been freely accepted within the society of states.

The same is true of the *in bello* rule that allows soldiers to undertake the duty of obedience. The rule permits a soldier x to participate in a war that her state fights, without making sure that the war is just. It does not follow that participating in the war would have been permissible for x, had the rule not been accepted. Rather, we argued that the rule that permits obedience is fair and mutually beneficial merely because it makes armies a better tool for addressing aggression and for deterring potential aggressors. We inferred that soldiers' obedience is permissible—or more accurately, that soldiers have a liberty-right to harm enemy combatants—merely because presumably, the rule that permits it has been freely accepted by states and by the individuals that they represent.

We conjecture that the prohibitive rules that the war agreement contains have Razian authority. They constrain the behaviour of states, soldiers, and civilians, in light of pre-contractual morality. In contrast, the permissive rules that the agreement contains are there in order to deter potential violators of these prohibitions. The parties accept them in the hope that they will not have to exercise the rights that these rules confer on them. We saw that deterrence can be achieved by (pre-contractually) disproportionate harm, or by allowing individuals not to respond to the moral reasons that apply to them. Hence, the permissive rules are morally effective merely because their addressees accept them. Absent their acceptance, the actions they permit might well be impermissible.

8.2 The Scope and Limitations of Contractarianism

The scope of our defence of the traditional war convention is limited in several respects. The most significant limitation concerns the convention's understanding of the empirical background in light of which the parties design the agreement; for instance, the assumption that the world is divided into mostly decent states, a minority of indecent states, a minority of very small and weak states, and a minority of decent and indecent stateless nations (some of which are entitled to political independence in their own states). The account further assumes a set of empirical assumptions regarding the motives of soldiers, leaders, and civilians. We do not pretend to have fully established these assumptions. In making assessments about fundamental aspects of human nature and of the relations between peoples and states, empirical data can take us only so far. Neither can the assessment that it is possible to reform the international regime in a way that will reduce the number of wars be drawn from a set of well-established empirical premises, nor can the opposite view (that such reform is unrealistic and might be counterproductive).

Another empirical assumption regards the actual acceptance of the war agreement. The actual compliance of international actors with the war convention is somewhat weak. In too many cases, claims of self-defence cover other motives for war that are clearly unwarranted by the agreement. Neither the UN nor individual states are able to enforce efficiently the 2(4) prohibition on use of force. We assume nevertheless that the *ad bellum* agreement improves the way wars are fought in comparison with 'the state of nature', namely, the state of affairs in which there is no agreement about the rules for waging and conducting war. We believe that despite its weak compliance and weak enforceability, the war agreement is still (*ex ante*) mutually beneficial for all parties, and that by virtue of this fact, the parties freely accept it.

Moreover, the factual background that the war agreement presumes might be changing in a way that will eventually undermine its moral standing. Imagine an international community that is governed not only by states and nations, but also by multinational corporations. Imagine further that while the latter have no military force, they are richer than states and have more 'soft power'. Under many circumstances, these corporations might shape the legal system in a way that suits their interests. Finally, imagine that public opinion is organized through social media, and that, as a result, in many fields, individuals are not represented by the state in which they reside, but by the companies whose services they consume. Against this background, the contractual deals that are signed by state governments may not reflect the interests of all relevant actors.

The empirical background might be changing in another regard as well. As we suggest reading it, the parties to the war agreement assume that in most conflicts, it would be *ex ante* better for the parties to bargain rather than to fight. But this might change. The technological superiority of some parties may render the anarchical society of states *asymmetrical*, in which case undertaking the duty not to use force in enforcing pre-contractual rights might not be beneficial to strong parties.

Having said that, it is important to emphasize that contractarianism challenges revisionism, whatever the empirical background may be. While revisionists believe that the morality of killing in war can be copied from the (pre-contractual) morality of individual self-defence, we argue that it cannot. To determine the moral norms that regulate war, one must inquire into the existing rules concerning war and determine whether they are good enough. The quality of the rules is independent of the morality of self-defence. If the rules that govern war are sufficiently good, they might be morally binding even if they cannot be deduced from individual morality of self-defence.

Another set of limitations involves the fact that, like all legal systems, the war agreement necessarily leaves lacunae. The legislators—viz., the states that entered the contract—could respond only to situations that they could have imagined at the time they entered the agreement. For example, cyber wars are not regulated by any cyber war agreement.

This absence of legal rules to regulate cyber war might have two opposing explanations. The discouraging (intuitive) explanation is that problems of coordination and mistrust prevent states from reaching agreement on rules to initiate, conduct and end cyber wars. The encouraging explanation is that cyber wars ought not to be regulated by legal rules. Why not? Because decent states ought to subject themselves to such an agreement only if they can know *ex ante* that following its rules is typically in their best interest and only if doing so does not solidify or create injustice. But legislating a system of rules that applies under only few and very select circumstances would be too costly. Legislating a system of rules that regulate conduct under circumstances in which it is in the interest of the parties to deliberate on the merits of the case is inappropriate. According to the encouraging explanation, there is no cyber war agreement for the simple reason that when it comes to cyber war, pre-contractual morality is our best guide. Hence, in the case of cyber wars, states ought to consider each case as it occurs.

Consider another lacuna. The war agreement regards combatants—employees of states or of non-state actors—as the main agents of war; thus, it excludes mercenaries—private individuals or groups whose fighting services are subcontracted by the state or by some parallel actor to assist in its war activities. When mercenaries are placed directly under the command chain of the state (namely, its army), there is no reason to think that they are excluded from the regular *in bello* rules that apply to all soldiers. However, as often is the case, their employment may be indirect and clandestine, making it unclear whether they are included in the war agreement. The outsourcing of war is left unregulated. The encouraging explanation is that there is no good set of rules that states can form in advance to govern the appropriate use of mercenaries. Alternatively, there are not sufficient cases of this type for the regulation to cover—hence states should not invest in regulating this issue.

The nature of the war agreement as a set of legal rules accounts for another important limitation. No set of legal rules can prescribe correct conduct in all situations.[1] For example, in most contexts, the set of rules to which drivers are subject must include an imposed speed limit. Still, a small number of especially bad drivers would drive carelessly without violating the high speed rule, whereas a small number of especially good drivers would drive carefully while violating the speed limit rule. Likewise, there may be cases in which violation of the contractual duties undertaken by a state is the right thing to do. Despite the fact that by accepting the war agreement, states and individuals undertake certain duties, in exceptional cases they ought to violate them.

Wars are so deeply ingrained in human history and civilization[2] that one cannot realistically hope that the day will come when they will be abolished. Therefore, the international community has a duty to regulate wars in the most humane and rights-respecting manner. Actually, it has already done so to a large extent in documents like the UN Charter and the Geneva Conventions. Our book should be seen as an attempt to explain why, correctly understood, these documents, by and large, are morally binding and why it is good that they are so. The limitations we listed above are numerous, but they are essential to any regulative project.

8.3 War and Tragedy

To say that wars are 'tragic' is almost a cliché (a quick online search yields more than 13 million results for the expression 'tragedy of war'). If all this expression means is events with harmful results to (innocent) human beings, then of course, wars are tragic—just like car accidents, earthquakes, and some diseases are. Applying the term 'tragic' in this very broad sense to wars would be uncontroversial, but would not illuminate anything particularly interesting about their moral nature. To contribute to such illumination, we should consider a narrower understanding of tragedy as the unavoidability of wrongdoing. What we have in mind are cases of moral dilemmas, understood as situations 'in which whatever

---

[1] See Paul Robinson, 'A Theory of Justification: Societal Harm as a Prerequisite for Criminal Liability', *UCLA L. Rev.* 23 (1975–1976), 271–2.
[2] See Azar Gat, *War in Human Civilization* (Oxford: Oxford University Press, 2008).

one does either one will wrong people, or one will fail in some binding duty... It is the essence of dilemmas that those facing them have no morally acceptable option.'[3] Such situations are rightly referred to as *moral* tragedies, a usage that preserves the classic use of the concept: (a) the agent unwillingly finds himself in a state of affairs in which he is doomed to do wrong; (b) nothing can prevent this moral fall; and (c) it threatens the very existence of the agent—whatever he does 'would irreparably damage one of the projects or relationships which he pursued and which shapes his life'.[4]

Are wars moral tragedies in this narrower sense? The unjust side to war obviously does wrong, but the wrong is not unavoidable; it can and should be avoided. So if the claim has any bite, it refers to the just side, by which we mean the side that is justified in going to war and that makes a sincere effort to follow the *in bello* rules. Can politicians and soldiers of the just side nonetheless be said 'to wrong people or to fail in some binding duty' (to use Raz's formulation)? To recap, nobody disputes the regrettable character of the results caused by war; the destruction, the suffering, the losses. However, on a smaller scale, such regrettable results are typical of other social institutions as well, such as punishment—think of the effects that incarcerating a criminal often have on his young children—yet no one would suggest that for these reasons, any instance of imprisoning criminals is a case of moral tragedy. Therefore, the question facing us is not whether or not war brings about very unwelcome results—of course it does—but whether it involves *unavoidable wrongdoing*.

Some philosophers answer this question in the positive, mainly on the basis of the almost inevitable—albeit unintentional—harm to civilians, many of whom 'have nothing or very little to do with the aggression that instantiated a justified response by some victim (or some third-party defender): in particular, people who are neither morally responsible for the aggression nor making any causal contribution to it'.[5] Maiming and killing such people seems 'substantially unjust'[6] and a clear case of wrongdoing. Note that this position does not entail pacifism. The point is that even when wars are morally justified overall, they still involve unavoidable wrongdoing; hence the moral tragedy.

This view seems to be especially true for revisionists. For such revisionists, not only are many civilians morally innocent (hence wronged when attacked by Just Combatants), but also many combatants are—those who bear no responsibility for their country's aggression, and those who make no causal contribution (and at times a negative one) to its execution. Nonetheless, revisionists believe that some

---

[3] Raz, *The Morality of Freedom*, 359–60. Other philosophers think that dilemmas must also be irresolvable. For the difference between these views, see Daniel Statman, *Moral Dilemmas* (Amsterdam: Rodopi, 1995), 13–16.
[4] Raz, *The Morality of Freedom*, 366.
[5] Michael New, 'The Tragedy of Justified War', *International Relations* 27/4 (2013), 466.
[6] Ibid.

wars are not only legitimate but required; 'to fail to resist aggression is to somehow acquiesce morally to it in an unacceptable way',[7] hence the moral necessity of fighting. But fighting inevitably wrongs innocents; hence the tragic trap.

In contrast, the contractarian view of war denies that wars are moral tragedies in the sense just alluded to, and in particular that wars are tragedies even from the perspective of the just side. To be sure, when Just Combatants kill people who are not pre-contractually legitimate targets, that is very sad and regrettable, but it involves no violation of the latter's rights. The point is easy to see with regard to soldiers who, by their very conscription into the army, waive their right not to be attacked by enemy soldiers. But it applies to civilians as well, who authorize their states to waive in their name their right not to be harmed collaterally. Thus, insofar as Just Combatants fight in accordance with the accepted *in bello* rules, they violate no rights and hence are not trapped in a moral tragedy. Note that such violation cannot be ascribed to the politicians (of the just side) either. If they are justified in ordering their army to go to war, and if the war is fought by the rules, they cannot be said to be involved in bringing about a moral tragedy. By contrast, politicians of the unjust side *can* be said to be responsible for the shedding of innocent blood; yet in their case, there is no moral *tragedy* involved (in the sense of unavoidable wrongdoing). The unjust side can and morally should refrain from going to war.

This fundamental debate about the moral nature of war influences the emotional burden imposed upon soldiers. In our view, soldiers can be assured that if only they follow the *in bello* rules (which are intentionally drafted to be easily understood and easily implemented), they can emerge from the war morally clean, so to say. They will have seen painful sights and will have caused regrettable results, but their moral consciences could remain untouched. By contrast, in the revisionist perspective, the burden on soldiers is greater than they can bear, to use the famous biblical expression. When revisionists send soldiers to the battlefield, they almost assure them that they (the soldiers) will do wrong, either because they'll turn out to be fighting for the unjust side, or because they will intentionally kill enemy soldiers who are not liable to defensive killing, or because they will inevitably bring about the deaths of innocent civilians as a foreseeable side-effect of attacks on military targets (or because of some combination of the three). With this constant sense of sin, it is hard to see how, psychologically, conscientious soldiers could bring themselves to fight effectively—or even to fight at all. The only way to ease this burden would be to hide the moral doctrines of revisionism from them, but that would be a tactic that revisionists, eager to reform the reality of war, would find hard to accept. Regardless of the theoretical difficulties in revisionism

---

[7] Rodin, *War and Self-Defense*, 198. For a critical discussion of Rodin (and others') use of what has been termed the 'tragedy solution', see Daniel Statman, 'Moral Tragedies, Supreme Emergencies and National-Defense', *Journal of Applied Philosophy* 23/3 (2006), 311–22.

explicated in Chapter 1, it also imposes a psychological burden on soldiers that is unbearable and unfair.

8.4 Revisionism and Contractarianism

In Chapter 1, we indicated that despite their thorough criticism of traditional just war theory, individualists end up accepting its basic tenets. They propose all kinds of reforms in international law and international institutions.[8] But nevertheless they reluctantly concede that until these reforms are agreed upon and are implemented, it is morally acceptable—indeed, morally *desirable*—to stick to the current legal regime. Moreover, as mentioned briefly in the Introduction,[9] sometimes, when revisionists try to justify this move, they too, rather surprisingly, rely on contractarian arguments. Here is one illustration from McMahan. Towards the end of his 2004 seminal paper, after having explained at length the moral flaws in the current war convention, he offers the following observation:

> Thus far in this article I have focused on what I will refer to as the "deep" morality of war... But there is another dimension to the morality of war that I have not explored: the laws of war, which are conventions established to mitigate the savagery of war... Given that general adherence to certain conventions is better for everyone, all have a moral reason to recognize and abide by these conventions. For it is rational for each side in a conflict to adhere to them only if the other side does. Thus if one side breaches the understanding that the conventions will be followed, it may cease to be rational or morally required for the other side to persist in its adherence to them. A valuable device for limiting the violence will thereby be lost, and that will be worse for all.[10]

We ourselves could have written this paragraph. It assumes the idea of mutual benefit ('general adherence to the war convention is better for everyone'), some notion of mutuality ('it is rational for each side in a conflict to adhere to them only if the other side does') and some idea of fairness ('if one side breaches the understanding that the conventions will be followed, it may cease to be rational

---

[8] See, for instance, Rodin, *War and Self-Defense*, who proposes the idea of 'an ultra-minimal state' which would hold a monopoly on the use of military force, together with a minimal judicial mechanism for the resolution of international and internal disputes (187) and our discussion of this proposal in section 3.6.
[9] Circa fn. 10.
[10] McMahan, 'The Ethics of Killing in War', 730. In a similar vein, see his argument about the normative results of violations of the war convention by one's enemy: 'When the different parties agree to abide by a certain rule because universal compliance is better for all than compliance by neither, the violation of the rule by one side may have the effect of releasing the other from its duty of compliance' (McMahan, 'The Morality of War and the Law of War', 35).

or morally required for the other side to persist in its adherence to them' [emphases added]). McMahan seems to realize that the only way to save the war convention from his critique is by some kind of contractarian move.

While it is a bit surprising to find this move in McMahan, it is more natural to find it in Walzer, who says the following:

> The moral reality of war can be summed up in this way: when combatants fight freely, choosing one another as enemies and designing their own battles, their war is not a crime...[The rules that govern their military conduct] rest *on mutuality and consent.*[11]

Thus, ultimately, the two central opponents in the current debate on just war theory agree that neither consequentialism nor right-based morality is sufficient to make sense of the rules of war, and that we must rely on some kind of contract-based morality. What we have tried to do in this book is to develop this insight at length; to clarify its theoretical commitments, explicate its implications to the different levels of (the morality of) war, and provide an initial justification.

McMahan might respond by saying that we do not give due weight to the distinction between the deep morality of war and the laws of war (see the above citation). While there may be utilitarian and perhaps contractarian reasons to comply with the latter, they do not touch on the real moral issues—on the *deep* morality of war. But this distinction, which following McMahan has become standard in revisionist writing,[12] seems to us misleading. Let us briefly explain why.

First, notably, as a technical term, 'deep morality' is unknown outside the ethics of war, and even within that field, unknown prior to the work of McMahan.[13] This limited use of a term that would seem to apply to all areas of applied ethics does not by itself constitute a reason against it, but it does raise some suspicion with regard to its theoretical underpinning.

Second, presenting the distinction as one between the (deep) morality of war and the *laws* of war is misleading in any case, as if there were a conflict here between what is required by morality and what the law—probably on some positivist understanding—requires. But if that were the case, then the moral guidance would be unequivocal; namely, to follow revisionist advice, though all kinds of non-moral considerations might pull in a different direction. What

---

[11] Walzer, *Just and Unjust Wars*, 37, emphasis added.
[12] See, for instance, Fabre, *Cosmopolitan War*, 168 ('here as throughout the book, I seek to inquire into the deep morality of war, not (or at least not primarily) its legality of the desirability of institutionalizing its constituent rights'); James Pattison, 'Deeper Objections to the Privatisation of Military Force', *Journal of Political Philosophy* 18/4 2010, 433, n. 33; and Haque, 'Law and Morality at War', 81 ('When I refer to "the morality of war" I will be referring to the deep morality of war in McMahan's sense').
[13] A Google Scholar search for the string 'deep morality of war' yields more or less the same results as the string 'deep morality' + McMahan.

McMahan really has in mind is a distinction between two types or two levels of moral considerations; those that are in some sense 'deep', probably having to do with fundamental human rights, and those that are not deep—maybe *shallow* is the appropriate term—mainly involving consequences. He concedes that there are *moral* reasons to comply with the current war convention, though they are, in a sense, shallow. As far as the deep morality of war is concerned, such compliance is nonetheless problematic.

However (and now we turn to our third and main criticism), the deep versus shallow distinction creates the misleading impression that one should give priority to considerations stemming from deep morality over those stemming from shallow morality. But, clearly, even revisionists do not assume such a priority—neither in general, nor in the field of war. To the contrary, McMahan's point in the passage cited above was precisely that although deep morality would often require diverting from the current legal regime, other considerations (which we proposed to see as contractarian) mandate compliance with it, and it is these *latter* considerations that take precedence, not the former.[14] The broader theoretical point is that if rights-based reasons are sometimes overridden by other moral reasons—which very few thinkers would deny—then saying that the former stand for 'deep' morality while the latter for 'shallow' can only create confusion.

We note in passing that a similar difficulty applies to the 'in principle' rhetoric that seems to play a role similar to the one played by the notion of deep morality. In this vein, it is often said that although some course of action is (or could be) justified in principle, for pragmatic reasons it ought not to be taken.[15] Again, this terminology conveys the misleading impression that what is 'in principle' justified is also justified overall, or that what is in principle forbidden is also forbidden overall—which is simply not the case. To the contrary, statements such as, 'in principle, *a* would be permissible' are almost always followed by a reservation stating that *a* is impermissible overall; likewise with 'in principle forbidden', which is typically followed by reservations along the lines of, 'but, given such and such circumstances, *a* is actually *not* forbidden'.

Back to our main point, then. We propose that the distinction between the deep morality of war and the laws of war be jettisoned altogether. When some moral decision involves conflicting considerations, it is best analysed in terms of the good old distinction between what one ought to do *prima facie* (or *pro tanto*) and

---

[14] For a similar criticism, see Nathanson, *Terrorism and the Ethics of War*, 297–8.
[15] See, for instance, Fabre, *Cosmopolitan War*, 105. Fabre argues that although on principle, subsistence wars might be morally justified (i.e. they might provide a just cause for war), there are powerful considerations against them. For a more detailed discussion of Fabre's view on this, see Daniel Statman, 'Fabre's Crusade for Justice: Why We Should Not Join' *Law and Philosophy* 33 (2014), 337–60. See also Jeff McMahan himself, who argues in 'Torture in Principle and in Practice', *Public Affairs Quarterly*, 22/2 (2008) that although torture can be morally permissible in principle, this fact is 'virtually irrelevant in practice' and 'it is morally necessary that the law, both domestic and international, should prohibit the practice of torture absolutely—that is, without exceptions' (111).

what one ought to do *overall*, or *all-things-considered*, rather than in terms of the distinction between deep and non-deep morality. All moral considerations initially claim to be behavioural guides, and to treat some considerations as 'deeper' than others is normatively empty. Depending on the circumstances, consideration C1 might override C2 in situation S1, but give way in situation S2. Therefore—to get back to revisionism and the ethics of war—once revisionists acknowledge that consequentialist and contractarian considerations are morally relevant, and, moreover, that they can override rights-based considerations, there is no point in saying that the latter are 'deeper' than the former.

8.5 Last Word

Wars are among the most irrational institutions of human society. In almost all cases, negotiation and compromise would achieve much better results for all sides involved. What's worse is that the sides often know this, but nevertheless find themselves shedding each other's blood and wreaking terrible destruction and misery. The acknowledgement of this irrationality led thinkers in the age of the Enlightenment to assume that as human beings grow more enlightened, 'there will be no war, no crimes, no administration of justice as it is called, and no government'.[16]

The wars of the twentieth century undermined this optimistic picture and put an end to the illusion that institutional reforms in the international arena, such as the establishment of the League of Nations and later of the United Nations, could guarantee world peace. The existence of evil individuals and of rogue states provides part of the explanation for why war cannot be eradicated. Benign self-interest coupled with epistemic shortcomings and constant suspicion of others also seen as self-interested and epistemically deficient is another source of violence.[17]

The link between war and such inherent features of human nature might explain why world peace is such a dominant theme in prophesies about the end of times, like the well-known one in Isaiah that refers to a time when nations will beat their swords into plowshares and their spears into pruning hooks (Isaiah 2, 2). Because the complete avoidance of war necessitates a fundamental transformation of

---

[16] William Godwin, *An Inquiry Concerning Political Justice* (London: G.G.J. and J. Robinson, 1793), 871.

[17] In our profound skepticism about the possibility of eradicating war, we come close to Nigel Biggar's view who sees himself as 'a realist about the fact of intractable human vice on the international stage' (*In Defense of War* [New York: Oxford University Press, 2013], 11) which makes war necessary to protect peace. Biggar contends that those who think that international conflicts can always be prevented by other means suffer from 'the virus of wishful thinking'.

human nature, it can be achieved only in some miraculous reality marking the end of human history as we know it.

Until this Messianic reality comes into existence, states would be better off agreeing on rules to regulate war that would, on the one hand, facilitate effective self-defence, while on the other, reduce the killing and harm they cause. A version of these rules has already been agreed upon and inscribed in central international documents. They are acknowledged by most states and shape the global public debate about war. What is needed for wars to be more humane is not to undermine these rules or to replace them by a completely new system, but to strengthen the commitment to them among states, NSAs, and fighting forces. We hope that this book makes a modest contribution to the achievement of this goal.

Bibliography

Abrahms, Max, 'Does Terrorism Really Work? Evolution in the Conventional Wisdom since 9/11', *Defence and Peace Economics* 22 (2011), 583-94.

Anderson, Elizabeth, 'What is the Point of Equality?' *Ethics* 109/2 (1999), 287-337.

Anderson, Kenneth and Matthew C. Waxman, 'Law and Ethics for Autonomous Weapon Systems: Why a Ban Won't Work and How the Laws of War Can', American University Washington College of Law. Research Paper No. 2013-11. Available at http://papers.ssrn.com/sol3/papers.cfm?abstract_id=2250126.

Angstrom, Jan and Isabelle Duyvesteyn (eds.), *Understanding Victory and Defeat in Contemporary War* (New York: Routledge, 2007).

Applbaum, Arthur I., *Ethics for Adversaries: The Morality of Roles in Public and Professional Life* (Princeton, NJ: Princeton University Press, 1999).

Ariely, Dan, *The (Honest) Truth About Dishonesty: How We Lie to Everyone—Especially Ourselves* (New York: HarperCollins, 2012).

Arneson, Richard J., 'Equality and Equality of Opportunity for Welfare', *Philosophical Studies* 55 (1989), 77-93.

Arneson, Richard J., 'Luck-Egalitarianism and Prioritarianism', *Ethics* 110 (2000), 339-49.

Arneson, Richard J., 'Just Warfare Theory and Noncombatant Immunity', *Cornell International Law Journal* 39 (2006), 663-88.

Arreguín-Toft, Ivan, *How the Weak Win Wars: A Theory of Asymmetric Conflict* (New York: Cambridge University Press, 2005).

Ashford, Elizabeth and Tim Mulgan, 'Contractualism', *Stanford Encyclopedia of Philosophy*, 2012.

Avraham, Ronen, 'Private Regulation', *Harvard Journal of Law and Public Policy* 34/2 (2011) 546-638.

Bazerman, Max H. and Ann E. Tenbrunsel, *Blind Spots: Why We Fail to Do What's Right and What to Do about It* (Princeton, NJ: Princeton University Press, 2011).

Bazerman, Max H., Ann E. Tenbrunsel, and Kimberly Wade-Benzoni, 'Negotiating with Yourself and Losing: Making Decisions with Competing Internal Preference', *Academy of Management Review* 23 (1998), 225-41.

Beitz, Charles R., 'Bounded Morality: Justice and the State in World Politics' *International Organization* 33/3 (1979), 405-24.

Beitz, Charles R., 'Nonintervention and Communal Integrity', *Philosophy and Public Affairs* 9 (1980), 385-91.

Bellamy, Alex J., 'Supreme Emergencies and the Protection of Noncombatants in War' *International Affairs* 80/5 (2004), 829-50.

Bellamy, Alex J., *Massacres and Morality: Mass Atrocities in an Age of Civilian Immunity* (New York: Oxford University Press, 2012).

Benbaji, Yitzhak, 'Culpable Bystanders, Innocent Threats and the Ethics of Self-Defense', *Canadian Journal of Philosophy* 35 (2005), 585-622.

Benbaji, Yitzhak, 'The Responsibility of Soldiers and the Ethics of Killing in War', *The Philosophical Quarterly* 57 (2007), 558-73.

Benbaji, Yitzhak, 'A Defense of the Traditional War-Convention', *Ethics* 118 (2008), 464-95.

Benbaji, Yitzhak, 'The War Convention and the Moral Division of Labor', *Philosophical Quarterly* 59 (2009), 593–618.
Benbaji, Yitzhak 'Contractarianism and Emergency' in Hanoch Sheinman (ed.) *Understanding Promises and Agreements: Philosophical Essays* (New York: Oxford University Press, 2010), 342–65.
Benbaji, Yitzhak, 'The Moral Power of Soldiers to Undertake the Duty of Obedience', *Ethics* 122 (2011), 43–73.
Benbaji, Yitzhak, 'Against a Cosmopolitan Institutionalization of Just War' in Yitzhak Benbaji and Naomi Sussmann (eds.), *Reading Walzer: Sovereignty, Culture and Justice* (New York: Routledge, 2013), 233–56.
Benbaji, Yitzhak, 'Justice in Asymmetric Wars: A Contractarian Analysis', *Law and Ethics of Human Rights* 6 (2013), 157–83.
Benbaji, Yitzhak, 'A Contractarian Account of the Crime of Aggression', in Cécile Fabre and Seth Lazar (eds.), *The Morality of Defensive War* (Oxford: Oxford University Press, 2014), 159–84.
Benbaji, Yitzhak, 'Legitimate Authority in War' in Helen Frowe and Seth Lazar (eds.), *Oxford Handbook of Ethics of War* (Oxford: Oxford University Press, 2018), 294–314.
Benbaji, Yitzhak, 'State Self-Defense' in Larry May et al. (eds.), *Cambridge Handbook of Just War* (Cambridge: Cambridge University Press, 2018), 59–79.
Benvenisti, Eyal, 'Human Dignity in Combat: The Duty to Spare Enemy Civilians', *Israel Law Review* 39/2 (2006), 81–109.
Biddle, Stephen, *Military Power: Explaining Victory and Defeat in Modern Battle* (Princeton, NJ: Princeton University Press, 2004).
Biggar, Nigel, *In Defence of War* (New York: Oxford University Press, 2013).
Blum, Gabriella, 'The Dispensable Lives of Soldiers', *Journal of Legal Analysis* 2/1 (2010), 69–124.
Blum, Gabriella, 'On a Differential Law of War', *Harvard International Law Journal* 52/1 (2011), 163–218.
Bond, Brian, *The Pursuit of Victory: From Napoleon to Saddam Hussein* (Oxford: Oxford University Press, 1996).
Brennan, Geoffrey, Lina Eriksson, Robert E. Goodin, and Nicholas Stouthwood, *Explaining Norms* (Oxford: Oxford University Press, 2013).
Bull, Hedley, *The Anarchical Society: A Study of Order in World Politics* (Oxford: Oxford University Press, 1977).
Burri, Susanne, 'The Toss-Up Between a Profiting, Innocent Threat and His Victim', *The Journal of Political Philosophy* 23 (2015), 146–65.
Byers, Michael and Simon Chesterman, 'Changing the Rules about Rules? Unilateral Humanitarian Intervention and the Future of International Law', in J. L. Holzgrefe and Robert O. Keohane (eds.), *Humanitarian Intervention: Ethical, Legal and Political Dilemmas* (Cambridge: Cambridge University Press, 2003), 177–203.
Byman, Daniel and Taylor Seybolt, 'Humanitarian Intervention and Communal Civil Wars', *Security Studies* 13 (2003), 33–78.
Coady, Cecil A. J., 'Terrorism, Morality and Supreme Emergency', *Ethics* 114 (2004) 772–89.
Coady, Cecil A. J., 'Terrorism and Innocence', *The Journal of Ethics* 8 (2004), 37–58.
Cohen, Gerald A., 'On the Currency of Egalitarian Justice', *Ethics* 99 (1989), 906–44.
Cohen, Gerald A., 'Where the Action Is: On the Site of Distributive Justice', *Philosophy and Public Affairs* 26/1 (1997), 3–30.

Cohen, Marshall, 'Moral Skepticism and International Relation', in Charles Beitz, Marshall Cohen, Thomas Scanlon, and A. John Simmons (eds.), *International Ethics* (Princeton, NJ: Princeton University Press, 1985), 3-51.

Cole, Darrell, 'Death before Dishonor or Dishonor before Death? Christian Just War, Terrorism, and Supreme Emergency', *Notre Dame Journal of Law, Ethics and Public Policy* 16 (2002), 81-99.

Cross, Ben, 'Moral Philosophy, Moral Expertise, and the Argument from Disagreement', *Bioethics* 30 (2016), 188-94.

Dare, Tim, 'Robust Role Obligations: How Do Roles Make a Moral Difference', *Journal of Value Inquiry* 50 (2016), 703-19.

Davis, Nancy, 'Abortion and Self-Defense', *Philosophy and Public Affairs*, 13 (1984), 175-207.

Dill, Janina, and Henry Shue, 'Limiting the Killing in War: Military Necessity and the St. Petersburg Assumption', *Ethics and International Affairs* 26/3 (2012), 311-33.

Dinstein, Yoram, *War, Aggression and Self-Defence* (Cambridge: Cambridge University Press, 2001).

Doppelt, Gerald, 'Walzer's Theory of Morality in International Relations', *Philosophy and Public Affairs* 8/1 (1978), 3-26.

Dougherty, Tom, 'Moral Indeterminacy, Normative Powers and Convention', *Ratio* 29/4 (2016), 448-65.

Doyle, Michael W., *Striking First: Preemption and Prevention in International Conflict* (Princeton, NJ: Princeton University Press, 2008).

Duyvesteyn, Isabelle, 'Some Conclusions', in Jan Angstrom and Isabelle Duyvesteyn (eds.), *Understanding Victory and Defeat in Contemporary War* (New York: Routledge, 2007), 224-35.

Dworkin, Ronald, *Taking Rights Seriously* (London: Duckworth, 1977).

Dworkin, Ronald, 'What Is Equality, Part 1: Equality of Welfare', *Philosophy and Public Affairs* 10 (1981), 185-246.

Eck, David, 'Social Coordination in Scientific Communities', *Perspectives on Science* 24/6 (2016), 770-800.

Edmundson, William A., 'Rethinking Exclusionary Reasons', *Law and Philosophy* 12 (1993), 329-43.

Fabre, Cécile, 'Guns, Food, and Liability to Attack in War', *Ethics* 120/1 (2009), 36-63.

Fabre, Cécile, *Cosmopolitan War* (Oxford: Oxford University Press, 2012).

Fabre, Cécile, 'War Exit', *Ethics* 125/3 (2015), 631-52.

Fearon, James, 'Rationalist Explanations for War', *International Organization* 49 (1995), 379-414.

Fletcher, George P., *Romantics at War: Glory and Guilt in the Age of Terrorism* (Princeton, NJ: Princeton University Press, 2002).

Fletcher, George P. and David Ohlin Jens, *Defending Humanity: When Force is Justified and Why* (New York: Oxford University Press, 2008).

Frowe, Helen, 'Equating Innocent Threats and Bystanders', *Journal of Applied Philosophy* 25 (2008), 277-90.

Frowe, Helen, *Defensive Killing* (Oxford: Oxford University Press, 2014).

Gans, Chaim, 'Mandatory Rules and Exclusionary Reasons', *Philosophia* 15 (1986), 373-96.

Gardam, Judith G., 'Proportionality and Force in International Law', *American Journal of International Law* 87 (1993), 391-413.

Gat, Azar, *The Origins of Military Thought: From the Enlightenment to Clausewitz* (New York: Oxford University Press, 1989).
Gat, Azar, *War in Human Civilization* (Oxford: Oxford University Press, 2008).
Gauthier, David, *Morals by Agreement* (Oxford: Clarendon Press, 1986).
Gibbard, Allan, 'Natural Property Rights', *Nous* 10 (1976), 77–86.
Gilbert, Margaret, 'Agreements, Coercion, and Obligation', *Ethics* 103/4 (1993), 679–706.
Godwin, William, *An Inquiry Concerning Political Justice* (London: G.G.J. and J. Robinson, 1793).
Greene, Joshua, *Moral Tribes: Emotion, Reason, and the Gap Between Us and Them* (London: Atlantic Books, 2015).
Griffin, James, *On Human Rights* (Oxford: Oxford University Press, 2008).
Gross, Michael, *Moral Dilemmas of Modern War: Torture, Assassination, and Blackmail in an Age of Asymmetric Conflict* (New York: Cambridge University Press, 2009).
Gross, Michael, *The Ethics of Insurgency: A Critical Guide to Just Guerrilla Warfare* (Cambridge: Cambridge University Press, 2015).
Gross, Samuel R., Barbara O'Brien, Chen Hu, and Edward H. Kennedy, 'Rate of False Conviction of Criminal Defendants Who Are Sentenced to Death', *Proceedings of the National Academy of Sciences in the United States* 111/20 (2014), 7230–5.
Haque, Adil A., 'Law and Morality at War', *Criminal Law and Philosophy* 8 (2014), 79–97.
Haque, Adil A., *Law and Morality at War* (Oxford: Oxford University Press, 2017).
Harbom, Lotta and Peter Wallansteen, 'Armed Conflict and Its International Dimensions, 1946-2004', *Journal of Peace Research* 42 (2005), 623–35.
Hardimon, Michael O., 'Role Obligations', *Journal of Philosophy* 91/7 (1994), 333–63.
Hart, Herbert L. A., *The Concept of Law* (Oxford: Oxford University Press, 1961/2012).
Hendrickson, David C., 'In Defense of Realism: A Commentary on Just and Unjust Wars', *Ethics and International Affairs* 11 (1997), 19–53.
Hobbs, Richard, *The Myth of Victory: What is Victory in War?* (Boulder, CO: Westview Press, 1979).
Hoffmann, Stefan-Ludwig (ed.), *Human Rights in the Twentieth Century* (New York: Cambridge University Press, 2011).
Holmes, Robert L., *On War and Morality* (Princeton, NJ: Princeton University Press, 1989).
Hooker, Brad, *Ideal Code, Real World: Rule-Consequentialist Theory of Morality* (Oxford: Oxford University Press, 2000).
Hurka, Thomas, 'Proportionality in the Morality of War', *Philosophy and Public Affairs* 33 (2005), 34–66.
Hurka, Thomas, 'Liability and Just Cause', *Ethics and International Affairs* 21/2 (2007), 199–218.
Jacobs, Frans, 'Reasonable Partiality in Professional Ethics: The Moral Division of Labor', *Ethical Theory and Moral Practice* 8 (2005), 141–54.
Jenkins, Brian M., 'The Future Course of International Terrorism', in Paul Wilkinson and Alasdair Stewart (eds.), *Contemporary Research on Terrorism* (Aberdeen: Aberdeen University Press, 1987), 581–9.
Johnson, Dominic and Dominic Tierney, 'Essence of Victory: Winning and Losing International Crises', *Security Studies* 13 (2003–2004), 350–81.
Johnson, Dominic and Dominic Tierney, *Failing to Win: Perceptions of Victory and Defeat in International Politics* (Cambridge, MA: Harvard University Press, 2006).
Johnson, Dominic and Dominic Tierney, 'In the Eye of the Beholder: Victory and Defeat in US Military Operations' in Jan Angstrom and Isabelle Duyvesteyn (eds.), *Understanding Victory and Defeat in Contemporary War* (New York: Routledge, 2007), 46–76.

Kaldor, Mary, *New and Old Wars: Organized Violence in a Global Era* (Stanford, CA: Stanford University Press, 1999).
Kalshoven, Frits and Liesbeth Zegveld, *Constraints on the Waging of War: An Introduction to International Humanitarian Law* (Cambridge: Cambridge University Press, 2011).
Kant, Immanuel, 'On a Supposed Right to Lie From Altruistic Motives' in Lewis. W. Beck (ed.) *Critique of Practical Reason and Other Writings in Moral Philosophy* (Chicago, IL: University of Chicago Press, 1949).
Kitcher, Philip, 'The Division of Cognitive Labor', *Journal of Philosophy* 87/1 (1990), 5–22.
Kutz, Christopher, 'The Difference Uniforms Make: Collective Violence in Criminal Law and War', *Philosophy and Public Affairs* 33/2 (2005), 148–80.
Kutz, Christopher, 'Fearful Symmetry', in David Rodin and Henry Shue (eds.) *Just and Unjust Warriors: The Moral and Legal Status of Soldiers* (Oxford: Oxford University Press, 2008), 69–86.
Lackey, Douglas, 'The Intentions of Deterrence' in Steven Lee and Avner Cohen (eds.), *Nuclear Weapons and the Future of Humanity: The Fundamental Questions* (Totowa, NJ: Rowman and Allanheld, 1984), 307–19.
Lackey, Douglas, *The Ethics of War and Peace* (New Jersey: Prentice-Hall, 1989).
Larsdotter, Kersti, 'Culture and Military Intervention', in Angstrom Jan and Duyvesteyn Isabelle (eds.), *Understanding Victory and Defeat in Contemporary War* (New York: Routledge, 2007), 206–23.
Lazar, Seth, 'War', *Stanford Encyclopedia of Philosophy*, 2016.
Lazar, Seth, 'Responsibility, Risk, and Killing in Self-Defense', *Ethics* 119/4 (2009), 699–728.
Lazar, Seth, 'The Responsibility Dilemma for Killing in War: A Review Essay', *Philosophy and Public Affairs* 38/2 (2010), 180–213.
Lazar, Seth, 'Liability and the Ethics of War: A Response to Strawser and McMahan', in Christian Coons and Michael Weber (eds.), *The Ethics of Self Defense* (Oxford: Oxford University Press, 2011), 292–304.
Lazar, Seth, 'Necessity in Self-Defense and War', *Philosophy and Public Affairs* 40/1 (2012), 3–44.
Lazar, Seth, *Sparing Civilians* (Oxford: Oxford University Press, 2015).
Lee, Steven P., *Ethics and War: An Introduction* (Cambridge: Cambridge University Press, 2012).
Lewis, David, *Convention* (Cambridge, MA: Harvard University Press, 1969).
Luban, David, 'Just War and Human Rights', *Philosophy and Public Affairs* 9/2 (1980), 160–81.
Luban, David, 'Intervention and Civilization: Some Unhappy Lessons of the Kosovo War', in Pablo de Greiff and Ciaran Cronin (eds.), *Global Justice and Transnational Politics* (Cambridge, MA: MIT Press, 2002), 79–115.
Luban, David, 'Preventive War', *Philosophy and Public Affairs* 32 (2004), 207–48.
Madison, Powers, 'Efficiency, Autonomy, and Communal Values in Health Care', *Yale Law and Policy Review* 10/2 (1992), 316–61.
Mandel, Robert, 'Reassessing Victory in Warfare', *Armed Forces and Security* 33 (2007), 461–95.
Margalit, Avishai and Joseph Raz, 'National Self-Determination', *The Journal of Philosophy* 87 (1990), 439–61.
Margalit, Avishai, *On Compromise and Rotten Compromises* (Princeton, NJ: Princeton University Press, 2010).
Marmor, Andrei, *Social Conventions* (Princeton, NJ: Princeton University Press, 2009).

Martel, William C., *Victory in War: Foundations of Modern Military Policy* (New York: Cambridge University Press, 2007).
Mavrodes, George I., 'Conventions and the Morality of War', *Philosophy and Public Affairs* 4/2 (1975), 117–31.
May, Larry, *War Crimes and Just War* (New York: Cambridge University Press, 2007).
McMahan, Jeff, 'Self-Defense and the Problem of the Innocent Attacker', *Ethics* 104 (1994), 252–90.
McMahan, Jeff, 'The Ethics of Killing in War', *Ethics* 114 (2004), 693–733.
McMahan, Jeff, 'The Basis of Moral Liability to Defensive Killing', Philosophical Issues 15 (2005), 386–405.
McMahan, Jeff, 'Just Cause for War', *Ethics and International Affairs* 19/3 (2005), 1–21.
McMahan, Jeff, 'Killing in War: A Reply to Walzer', *Philosophia* 34 (2006), 47–51.
McMahan, Jeff, 'On the Moral Equality of Combatants', *Journal of Political Philosophy* 14/4 (2006), 377–93.
McMahan, Jeff, 'Torture in Principle and in Practice', *Public Affairs Quarterly* 22/2 (2008), 111–28.
McMahan, Jeff, 'The Morality of War and the Law of War', in David Rodin and Henry Shue (eds.), *Just and Unjust Warriors: The Moral and Legal Status of Soldiers* (Oxford: Oxford University Press, 2008), 19–43.
McMahan, Jeff, *Killing in War* (Oxford: Oxford University Press, 2009).
McMahan, Jeff, 'Intention, Permissibility, Terrorism, and War', *Philosophical Perspectives* 23 (2009), 345–72.
McMahan, Jeff, 'Who is Morally Liable to be Killed in War', *Analysis* 71/3 (2011), 544–59.
McMahan, Jeff, 'Duty, Obedience, Desert, and Proportionality in War: A Response', *Ethics* 122 (2011), 135–67.
McMahan, Jeff, 'What Rights May be Defended By Means of War?' in Cécile Fabre and Seth Lazar (eds.), *The Morality of Defensive War* (New York: Oxford University Press, 2014), 115–56.
McMahan, Jeff, 'The Prevention of Unjust Wars' in Yitzhak Benbaji and Naomi Sussmann (eds.) *Reading Walzer* (Abingdon: Routledge, 2014) 233–55.
McMahan, Jeff, 'Proportionality and Time', *Ethics* 125/3 (2015), 696–719.
McPherson, Lionel, 'Innocence and Responsibility in War', *Canadian Journal of Philosophy* 34/4 (2004), 485–506.
Melzer, Nils, *Direct Participation in Hostilities Under International Humanitarian Law* (Geneva: ICRC, 2009).
Merom, Gil, *How Democracies Lose Small Wars* (New York: Cambridge University Press, 2003).
Miller, David, 'Responsibility and International Inequality in the Law of Peoples', in Rex Martin and David A. Reidy (eds.), *Rawls's Law of Peoples: A Realistic Utopia?* (Oxford: Blackwell, 2006), 191–206.
Mitchell, Andrew D., 'Does One Illegality Merit Another? The Law of Belligerent Reprisals in International Law', *Military Law Review* 170 (2001), 155–77.
Moellendorf, Darrel, 'Jus ex Bello', *Journal of Political Philosophy* 16 (2008), 123–36.
Moore, Margaret, 'Collective Self-Determination, Institutions of Justice, and Wars of National Defence', in Cécile Fabre and Seth Lazar (eds.), *The Morality of Defensive War* (Oxford: Oxford University Press, 2014), 185–202.
Moore, Michael S., 'Authority, Law and Razian Reasons', *S. California Law Review* 62 (1989), 866–7.
Nagel, Thomas, *Mortal Questions* (New York: Cambridge University Press, 1979).

Nash, Laura L. *Good Intentions Aside: A Manager's Guide to Resolving Ethical Problems* (Boston, MA: Harvard Business School Press, 1993).
Nathanson, Stephen, *Terrorism and the Ethics of War* (Cambridge: Cambridge University Press, 2010).
Neff, Stephen, *War and the Law of Nations* (Cambridge: Cambridge University Press, 2005).
New, Michael, 'The Tragedy of Justified War', *International Relations* 27/4 (2013), 461–80.
Norman, Richard, *Ethics, Killing and War* (Cambridge: Cambridge University Press, 1995).
Norman, Richard, 'War, Humanitarian Intervention and Human Rights', in Richard Sorabji and David Rodin (eds.), *The Ethics of War—Shared Problems in Different Traditions* (Aldershot: Ashgate Publishing, 2005), 191–208.
Oberman, Kieran, 'The Myth of the Optional War: Why States Are Required to Wage the Wars They Are Permitted to Wage', *Philosophy and Public Affairs* 43/4 (2015), 255–86.
O'Neill, Onora, 'Children's Rights and Children's Lives', *Ethics* 98 (1988), 445–63.
Orend, Brian, 'Just and Lawful Conduct in War: Reflections on Michael Walzer', *Law and Philosophy* 20/1 (2001), 1–30.
Orend, Brian, *The Morality of War* (Peterborough, ON: Broadview Press, 2006).
Otsuka, Michael, 'Killing the Innocent in Self-Defense', *Philosophy and Public Affairs* 23 (1994), 74–94.
Øverland, Gerhard, 'Contractual Killing', *Ethics* 115 (2005), 692–720.
Parks, W. Hays, 'Air War and the Law of War', *Air Force Law Review* 32 (1990), 1–226.
Pattison, James, 'Deeper Objections to the Privatisation of Military Force', *Journal of Political Philosophy* 18/4 (2010), 425–47.
Pattison, James, 'When Is It Right to Fight? Just War Theory and the Individual-Centric Approach', *Ethical Theory and Moral Practice* 16 (2013), 35–54.
Pattison, James, 'The Case for the Nonideal Morality of War: Beyond Revisionism versus Traditionalism in Just War Theory', *Political Theory* 46/2 (2018), 242–68.
Pogge, Thomas 'On the Site of Distributive Justice: Reflections on Cohen and Murphy', *Philosophy and Public Affairs* 29 (2000), 137–69.
Posner, Eric and Alan Sykes, 'Optimal War and *Jus Ad Bellum*', *Georgetown Law Journal* 93 (2005), 993–1015.
Posner, Eric, 'Human Rights, the Laws of War, and Reciprocity', *Law and Ethics of Human Rights* 6/2 (2012), 147–71.
Primoratz, Igor, 'The Morality of Terrorism', *Journal of Applied Philosophy* 14 (1997), 221–33.
Primoratz, Igor, 'State Terrorism and Counter Terrorism', in Igor Primoratz (ed.), *Terrorism: The Philosophical Issues* (Basingstoke: Palgrave, 2004), 113–27.
Quinn, Warren S., 'Actions, Intentions, and Consequences: The Doctrine of Double Effect', *Philosophy and Public Affairs* 18/4 (1989), 334–51.
Quong, Jonathan, 'Killing in Self-Defense' *Ethics*, 119 (2009), 507–37.
Rawls, John, *The Law of Peoples* (Cambridge, MA: Harvard University Press, 1999).
Raz, Joseph, 'Reasons for Action, Decisions and Norms', *Mind* 84 (1975), 481–99.
Raz, Joseph, *The Morality of Freedom* (New York: Oxford University Press, 1986).
Raz, Joseph, *Practical Reason and Norms* (Princeton, NJ: Princeton University Press, 1990).
Reglitz, Merten, 'Political Legitimacy Without a (Claim-) Right to Rule', *Res Publica* 21/3 (2015), 291–307.
Ripstein, Arthur, *Force and Freedom: Kant's Legal and Political Philosophy* (Cambridge, MA: Harvard University Press, 2009).

Robinson, Paul, 'A Theory of Justification: Societal Harm as a Prerequisite for Criminal Liability', *UCLA L. Rev.* 23 (1975-1976), 266-92.
Rodin, David, *War and Self-Defense* (New York: Oxford University Press, 2002).
Rodin, David, 'Terrorism without Intention', *Ethics* 114 (2004), 752-71.
Rodin, David, 'The Ethics of Asymmetric War', in Richard Sorabji and David Rodin (eds.), *The Ethics of War—Shared Problems in Different Traditions*, (Aldershot: Ashgate Publishing, 2005), 153-69.
Rodin, David, 'Two Emerging Issues of Jus Post Bellum: War Termination and the Liability of Soldiers for Crimes of Aggression' in Jann Kleffner and Carsten Stahn (eds.), *Jus Post Bellum: Reflections on a Law of Transition from Conflict to Peace* (The Hague: T.M.C. Asser Press, 2008), 53-75.
Rodin, David, 'Ending War', *Ethics and International Affairs* 25/3 (2011), 359-67.
Rodin, David and Henry Shue, 'Introduction', in David Rodin and Henry Shue (eds.), *Just and Unjust Warriors: The Moral and Legal Status of Soldiers* (Oxford: Oxford University Press, 2008), 1-18.
Ryan, Cheyney, 'Moral Equality, Victimhood and the Sovereignty Symmetry Problem', in David Rodin and Henry Shue (eds.), *Just and Unjust Warriors: The Moral and Legal Status of Soldiers* (Oxford: Oxford University Press, 2008), 131-52.
Ryan, Cheyney, 'Democratic Duty and the Moral Dilemmas of Soldiers', *Ethics* 122 (2011), 10-42.
Scanlon, Thomas, *What We Owe to Each Other* (Cambridge, MA: Harvard University Press, 1998).
Sepielli, Andrew, *Along an Imperfect Lighted Path: Practical Rationality and Normative Uncertainty* (PhD Dissertation, Rutgers University, 2010).
Shapiro, Scott, 'Authority', in Scott Shapiro and Julius Coleman (eds.) *The Oxford Handbook of Jurisprudence and Philosophy of Law* (Oxford: Oxford University Press, 2004), 382-440.
Sharkey, Noel, 'Saying "No!" to Lethal Autonomous Targeting', *Journal of Military Ethics* 9 (2010), 369-83.
Sharvit, Baruch Pnina and Noam Neuman, 'Warning Civilians Prior to Attack under International Law: Theory and Practice', *International Law Studies* 87 (2011), 359-412.
Shue, Henry, 'War', in Hugh LaFollette (ed.) *The Oxford Handbook of Practical Ethics* (Oxford: Oxford University Press, 2003), 734-61.
Simmons, A. John, 'External Justifications and Institutional Roles', *Journal of Philosophy* 93 (1996), 28-36.
Smilansky, Saul, 'When Does Morality Win?' *Ratio* 23 (2010), 102-10.
Smith, Matthew N., 'Terrorism as Ethical Singularity', *Public Affairs Quarterly* 24 (2010), 229-45.
Sofaer, Abraham, 'The US Decision not to Ratify Protocol I to the Geneva Conventions on the Protection of War Victims (con'd), The Rationale for the United States Decision', *American Journal of International Law* 82 (1987), 784-7.
Southwood, Nicholas and Lina Eriksson, 'Norms and Conventions', *Philosophical Explorations* 14/2 (2011), 195-217.
Spiekermann, Kai, 'Review of *Explaining Norms*', *Economics and Philosophy* 31/1 (2015), 174-81.
Statman, Daniel, *Moral Dilemmas* (Amsterdam: Rodopi, 1995).
Statman, Daniel, 'Supreme Emergencies Revisited', *Ethics* 117 (2006), 58-79.
Statman, Daniel, 'Moral Tragedies, Supreme Emergencies and National-Defense' *Journal of Applied Philosophy* 23/3 (2006), 311-22.

Statman, Daniel, 'On the Success Condition for Legitimate Self-Defense', *Ethics* 118/4 (2008), 659–86.
Statman, Daniel, 'Can Wars Be Fought Justly? The Necessity Condition Put to the Test', *Journal of Moral Philosophy* 8 (2011), 435–51.
Statman, Daniel, 'Fabre's Crusade for Justice: Why We Should Not Join', *Law and Philosophy* 33 (2014), 337–60.
Statman, Daniel, 'Ending War Short of Victory? A Contractarian View of Jus *Ex Bello*', *Ethics* 125 (2015), 720–50.
Statman, Daniel, 'Drones and Robots: On the Changing Practice of Warfare' in Seth Lazar and Helen Frowe (eds.), *The Oxford Handbook of Ethics of War* (Oxford University Press, online edition, 2015).
Statman, Daniel, Raanan Sulitzeanu-Kenan, Micha Mandel, Michael Skerker, and Stephen De Wijze, 'Unreliable Protection: Proportionality Judgments in War' (unpublished manuscript).
Stilz, Anne, 'Territorial Rights and National Defense' in Cécile Fabre and Seth Lazar (eds.), *The Morality of Defensive War* (Oxford: Oxford University Press, 2014), 203–29.
Stocker, Michael, *Plural and Conflicting Values* (Oxford: Clarendon Press, 1990).
Strawser, Bradley J., 'Walking the Tightrope of Just War', *Analysis* 71/3 (2011), 533–44.
Tadros, Victor, *The Ends of Harm* (Oxford: Oxford University Press, 2011).
Tadros, Victor, 'Unjust Wars Worth Fighting For', *Journal of Practical Ethics* 4/1 (2016), 52–78.
Tadros, Victor, 'Anarchic War' (unpublished manuscript).
Tantaros, Andrea, 'This is War, Not a Law Seminar: Obama is More Carter than Churchill', *New York Daily News*, 30 September 2010, available at http://www.nydailynews.com/opinion/war-law-seminar-obama-carter-churchill-article-1.443668.
Temkin, Larry, 'Inequality: A Complex, Individualistic, and Comparative Notion', *Philosophical Issues* 11 (2001), 327–52.
Temkin, Larry, 'Egalitarianism Defended', *Ethics* 113 (2003), 764–82.
Tesón, Fernando (ed.), *The Theory of Self-Determination* (Cambridge: Cambridge University Press, 2016).
Thoma, Johanna, 'The Epistemic Division of Labor Revisited', *Philosophy of Science* 82/3 (2015), 454–72.
Thomson, Judith J., 'The Trolley Problem', *The Yale Law Journal* 94/6 (1985), 1395–1415.
Thomson, Judith J., 'Self Defense', *Philosophy and Public Affairs* 20 (1991), pp. 283–310.
Thomson, Judith J., 'Turning the Trolley', *Philosophy and Public Affairs* 36 (2008), 359–74.
Toner, Christopher, 'Just War and the Supreme Emergency Exemption', *Philosophical Quarterly* 55 (2005), 545–61.
United States Department of Transportation National Highway Traffic Safety Administration, Fatality Analysis Reporting System (FARC) Encyclopedia, Accessed 27/02/2017, Available at: https://www-fars.nhtsa.dot.gov/Main/index.aspx.
Valentini, Laura, '"When in Rome, Do as the Romans Do": Respect, Positive Norms, and the Obligation to Obey the Law' (unpublished manuscript).
Von Clausewitz, Carl, *On War*, translated and edited by Michael Howard and Peter Paret (Princeton, NJ: Princeton University Press, 1976, originally published in 1832).
Waldron, Jeremy, *Torture, Terror and Trade-Offs: Philosophy for the White House* (Oxford: Oxford University Press, 2010).
Walzer, Michael, *Arguing About War* (New Haven, CT: Yale University Press, 2004).
Walzer, Michael, 'The Moral Standing of States: A Response to Four Critics', *Philosophy and Public Affairs* 9/3 (1980), 209–29.

Walzer, Michael, *Just and Unjust Wars* (New York: Basic Books, 2006).
Walzer, Michael, 'Response to McMahan's Paper', *Philosophia* 34 (2006), 43–5.
Walzer, Michael, 'Coda: Can the Good Guys Win?', *The European Journal Of International Law* 24/1 (2013), 433–44.
Wasserstrom, Richard, 'Review of Michael Walzer's *Just and Unjust Wars: A Moral Argument with Historical Illustrations*', *Harvard Law Review* 92/2 (1978), 536–45.
Wendt, Alexander, 'A Comment on Held's Cosmopolitanism', in Ian Shapiro and Casiano Hacker-Cordon (eds.), *Democracy's Edges* (Cambridge: Cambridge University Press, 1999), 127–33.
Whitman, James Q., *The Verdict of Battle: The Law of Victory and the Making of Modern War* (Cambridge, MA: Harvard University Press, 2012).
Williams, Bernard, 'Moral Luck' in Daniel Statman (ed.) *Moral Luck* (New York: SUNY Press, 1993), 35–55.
Xavier, Marquez, 'An Epistemic Argument for Conservatism', *Res Publica* 22 (2016), 405–22.
Young, Aaron, Humayun J. Chaudhry, Xiaomei Pei, Katie Halbesleben, Donald H. Polk, and Michael Dugan, A Census of Actively Licensed Physicians in the United States, *Journal of Medical Regulation* 101/2 (2015), 8–23.
Zohar, Noam, 'Collective War and Individualistic Ethics: Against the Conscription of "Self-Defense"', *Political Theory* 21/4 (1993), 606–22.
Zohar, Noam, 'Can a War Be Morally "Optional"?' *Journal of Political Philosophy* 4/3 (1996), 229–41.
Zohar, Noam, 'Should the Naked Soldier be Spared? A Review Essay of Larry May, *War Crimes and Just War*', *Social Theory and Practice* 34/4 (2008), 623–34.
Zupan, Dan, 'A Presumption of the Moral Equality of Combatants: A Citizen-Soldier's Perspective', in David Rodin and Henry Shue (eds.), *Just and Unjust Warriors: The Moral and Legal Status of Soldiers* (Oxford: Oxford University Press, 2008), 214–25.

Index

For the benefit of digital users, indexed terms that span two pages (e.g., 52–53) may, on occasion, appear on only one of those pages.

absolutism 14–15
acceptance (of social rules) 7–8, 14n.16, 38, 39n.4, 40, 44, 46–9, 53, 56–7, 62, 64, 66–72, 79–80, 84–6, 117, 124–7, 129–30, 132, 135, 137–8, 143–4, 146–7, 149–51, 162–3, 176–8, 181–3, 185–7, 189–91,
193 *see also*: consent
act-focused morality 7
actuality (as a condition for the moral effectiveness of social rules) 43–8, 68–9, 71, 85–6, 116–17, 120, 132, 163, 177–8, 181, 191
advance warning 154 *see also*: pre-warning
agent-regret 131
aggression (legal prohibition on/crime of) 1, 4, 71, 77–8, 88–9, 92–3, 95–6, 98–100, 103, 113, 115–16, 118, 120, 128, 141, 151, 182, *see also*: Article 2(4) of the UN Charter; peace
Al-Qaeda 8
anarchy
minimally just 7–8, 72–3, 80–1, 84, 90–1, 95–8, 105–6, 115–18, 131–2, 135–6, 140–1, 151, 156, 163, 166, 171, 182–4
symmetrical 7–8, 72–4, 76–7, 84, 90–1, 95–8, 100, 105–6, 116–18, 122, 131–2, 135–6, 140–2, 151, 156–7, 163, 166, 171, 182–6
Applbaum, Arthur I. 67n.52
Aquinas, Thomas
'just war view' 2n.4
asymmetric wars 142, 152–4, 156, 180, 187–8
see also: irregular wars
authority
legitimate authority 52–3, 58, 85, 124, 159–60, 189–90
see: jus ad bellum-> legitimate authority
Razian 52–3, 85, 124, 189–91
Service conception of 52–3, 68, 159–60
Autonomous Weapon Systems
see: robots

"balance of terror" 171, 176–8
see also: weapons of mass destruction
"bargaining range" 74–5, 80, 90

battle of Chotusitz 108
battle of Malplaquet 107–8
behavioural ethics 158–9
Biggar, Nigel 199n.16
Blum, Gabriella 146, 180n.33

casus belli 71–2, 77, 85, 88–9, 110, 112–13
see also: just cause for war; jus ad bellum
Clausewitz, Carl von
anihaltion as the proper end of war 102–6, 115, 185
civilians
civilian immunity 5, 12, 27–8, 34, 133–7, 140–2, 152–4, 156–7, 159–62, 167–70, 183, 188, 190
collateral harm
see: collateral harm/ collateral damage/ collateral killing
moral equality of
see: moral equality of noncombatants/ civilians, noncombatants
Cohen, Gerald A. 54–5
collateral harm/ collateral damage/ collateral killing 8, 12, 24, 26–7, 34–6, 99, 116, 129, 133–5, 142–51, 153, 156–7, 161–2, 183, 187, 189, 195
see also: civilians
combatants
moral equality of
see: moral equality of combatants
consent 3, 6–8, 33, 38, 48, 62–3, 65–6, 84–6, 93, 116n.1, 117, 124–7, 130, 150–1, 197
hypothetical 44
coerced 63, 126–7
tacit/implicit 44, 49, 70
see also: acceptance
consequentialism 67, 172, 197–9
rule–Consequentialism (RC) 37, 66–8
see also: utilitarianism
convention 3, 39n.4, 41–2, 44–5, 45n.15, 46nn.19–20, 47n.22, 48, 63–4, 100, 108–10, 125, 139, 163–6, 168n.9, 176, 179, 181–2, 196–7
conservation of rights violations 64–6, 186

Davis, Jeremy 19–20
deep morality of war 160, 179, 196–9
deontology/ deontological morality
 134–5, 172
 threshold deontology 172
decent states/indecent states 2–5, 52, 69, 72–6,
 79, 82–3, 88–92, 96–7, 105–7, 110–13, 115,
 118–20, 122, 124, 132, 135–7,
 139–41, 143, 145, 147, 151, 156–7,
 160, 163–4, 169–71, 174–6, 178–80,
 183, 185–9, 191–2
Doctrine of Double Effect (DDE) 25–6,
 134–5, 174
 manipulative/opportunistic/exploitative
 killing 25–6, 134–5, 143
 eliminative killing 134–5
 the distinction between intentional and
 foreseeable killing 24–6, 143, 150
Dougherty, Tom 39n.4, 49
Doyle, Michael W. 11n.7, 78n.13
Dworkin, Ronald
 external preferences 45

Fabre, Cécile
 just cause in jus ad bellum 18–19
 'Quota View' of proportionality 114
 on the moral responsibility of civilians 134n.9
 subsistence wars 17, 21n.42, 198n.14
fairness 31–2, 40, 182
 as a condition for the moral effectiveness of
 social rules 7, 43, 45–9, 51–7, 61–6,
 68–72, 80–6, 89–91, 116–17, 119–20,
 127, 130, 132, 135, 137–8, 140–2, 149–57,
 162, 165, 177, 181, 183–6, 188–90,
 196–7
Frowe, Helen 9n.4

Gat, Azar 109
Gauthier, David
 contractarianism 3n.5
 Hobbesian version of contractarianism 67
Geneva Conventions 1, 9, 12n.11, 34–5, 157,
 165, 193
Gilbert, Margaret 126–7
global state 92
Greene, Joshua 79, 118
Gross, Michael 89n.28, 153n.40
Grotius, Hugo 2, 116n.1

humanitarian intervention 17–18, 73, 88, 170–1,
 182–3
habitual obedience 38–40, 44, 46–9, 56–7,
 63, 67
Hardimon, Michael 53, 125, 150n.37

Hart, H.L.A.
 practice theory of rules 38
 the concept of social rules 38, 38n.3, 46–7
 internal point of view 38, 46–7, 49, 68–9
 theory of legality 69
Haque, Adil A. 9n.4, 23n.48, 112, 119n.6, 134,
 143n.22
Hobbes, Thomas
 Hobbesian contractarianism 3n.5, 66–7
 Hobbesian view of (pre-contractual)
 morality 38–9, 67–9, 179
Hooker, Brad 73, 164n.2
 on rule consequentialism 66–7
Hurka, Thomas 3, 112, 117, 144
Hume, David 79

IDF 121n.9
imperfect duties 39n.6
 see also: imperfect obligation
imperfect obligation 60
 see also: imperfect duties
Independence Thesis 10–11, 13, 16
Individualism 7, 10, 13–18, 20–1, 23–4, 26–9,
 33–8, 42, 57–9, 68, 114,
 122–3, 129–30, 180, 185–6, 196
 see also: reductionism; revisionism; 'purism'
innocent aggressors 161
innocent bystanders 161, 173
international humanitarian law (IHL) 143n.22, 146
Intifada x
irregular wars 180
 see also: asymmetric wars
ISIS 8, 169, 179
Israel 86n.23, 121n.9
intentional killing 6, 12–13, 25, 150

jus ad bellum 3–8, 10n.5, 10–11, 13, 15–20, 27,
 35–6, 71–2, 77, 79–80, 84–90, 94, 96–8,
 103–5, 110–15, 119–20, 122–3, 140–1, 153–4,
 156–8, 164–5, 169–70, 178, 180, 182–8, 190–1
 threats to territorial integrity 4–5, 11, 15–18,
 20–1, 71–3, 76–8, 84–5, 88–9, 91, 95,
 98–100, 110–11, 112–13, 117–18, 120,
 148–9, 164–5, 182, 184–5, 187–8
 necessity/last resort 11, 71–2, 88–9, 96, 111
 proportionality 11, 19, 71–2, 85n.21, 88–9, 96,
 99, 111–15
 probability of success 11, 99, 111–12
 legitimate authority 11, 97, 123–4
 rightful intention 11
 see also: Independence thesis

jus in bello 3
In bello necessity 135, 146–7, 165, 183

In bello proportionality 1, 112, 134–5,
  144–5, 147–9, 161, 187
  see also: Independence thesis
jus ex bello 5, 8, 98–100, 106, 110–12, 115, 122–3
  just aim 98–100, 111–12, 187–8
  proportionality 114–15
  Quota View 114–114
jus post bellum 5, 118–19
jus terminatio 98
  see also: *jus ex bello*
just cause 6–8, 12–14, 18–19, 35–6, 71–2, 76, 78,
  82, 84–5, 89–91, 96, 98–100, 110–11, 117,
  121, 124–5, 144, 148, 153–4, 157–8, 198n.14
  see also: casus belli; jus ad bellum
just war theory 2–3, 98, 100, 197
  traditional 2, 6, 10, 16, 20–1, 28, 36, 68–9, 103,
  180, 196
justice
  distributive justice 17, 53–5, 82, 84, 92,
  119–20
  justice-promoting wars 17, 20, 73, 80–1, 84–5,
  88–9, 96–7, 100
  minimal justice 80–1, 83, 85
  see also: anarchy->minimally just
  Rawls' theory of justice 53–5

Kant, Immanuel 69, 181
  on world government 92
Kellogg-Briand pact 9
  see also: pact of Paris

Lackey, Douglas 168n.9
The Law of Armed Conflict (LOAC) 2–3, 5,
  17–18, 145, 157–8
Lazar, Seth 9n.3, 28
  on the permission to kill Unjust
  Combatants 29
  on Civilian Immunity 34
Lee, Steven P. 137n.13
Lewis, David
  on conventions 39n.4, 46n.19, 47n.22
liability for defensive killing 24, 31
  see also: McMahan, Jeff, resposibility
Locke, John
  Lockean view of (pre-contractual)
  morality 38–9, 67–9, 179
  property rights as pre-contractual rights
  39–40
Luban, David 17, 17n.27, 78n.13
luck egalitarianism 31–2

Marmor, Andrei 39n.4
Mavrodes, George 3
McPherson, Lionel 9n.4

McMahan, Jeff 9–10, 28, 33–4
  asymmetrical regime proposal 120–2
  contractarian arguments 3, 196–7
  deep morality of war 196
  on the Doctrine of Double Effect 134–5
  jus ad bellum court (international court)
  92–5
  just cause 19, 98–100
  just aim 98–101
  deep morality of war vs. laws of war 14n.14,
  197–8
  on punishing Unjust Combatants 29,
  58n.37
  on proportionality in the case of unjust
  combatants 144
  liability for defensive killing 30–1, 173
  on obedience 120–1
  on equality of combatants and the waiver
  argument 124–5
  on the moral authority of the current
  arrangements 36, 196–7
  on reprisals 167n.8
  on statism of traditional just war theory 180
  quota view 114
Miller, David 17n.26
minimally just anarchy
  see: anarchy->minimally just
moral division of labor 51–7, 55n.33, 59–61,
  65, 67–8, 67n.52, 122–4, 128, 130–1, 186
moral equality
  of noncombatants 12–13, 116–17, 148, 161
  of combatants 12–13, 21–2, 116–17, 124–5,
  128, 132, 161, 166, 186
  at the level of states 88–9
  see also: symmetry thesis
moral motivation 79
moral tragedies 193–5

Nagel, Thomas 56n.35, 60n.40, 150–1
Nathanson, Stephen 168n.9
national liberation 34, 90–1, 156
  wars of 102
natural rights 37–9, 42, 47–8, 59, 61, 65–6, 69,
  108, 126, 130
  see also: 'pre-contractual' rights and duties
necessity
  ad bellum necessity see: *jus ad bellum*->
  necessity
  in bello necessity see: *jus in bello*-> necessity
  contractual 53, 85
  pre-contractual 2–3, 38–9, 73, 134
noncombatants
  distinction between combatants and
  noncombatants 1, 24, 46–7, 113–14, 116

noncombatants (*cont.*)
noncombatant immunity 1, 9n.4, 134
see also: civilians
non-state actors (NSAs) 74, 89–91, 152–7, 162, 169, 171, 184, 188 *see also*: stateless nations
Norman, Richard 16, 18, 139n.18
nuclear weapons 105n.19, 176–8
see also: weapons of mass destruction

occupation 87, 101, 171–2

pacifism 18–21, 27, 76, 138–9, 194
contingent pacifism 35–6
pact of Paris 1
see also: Kellogg-Briand pact
Pattison, James 28, 122–3
Peace 2–6, 95–6, 100, 106, 110–12, 199–200
a crime against 4, 6, 87, 89, 91, 98, 100, 112, 175–6
see also: Article 4(2); aggression
perfidy 129
political participation 92, 94–5
'pre–contractual' rights 2–4, 6–7, 38–40, 43, 49, 55, 67–70, 77, 85, 89, 112, 129, 139, 143–4, 151, 176, 179, 181–2, 184, 187–8, 192
see also: natural rights
pre-warning 156
see also: advance warning
preventive wars 4, 71, 73, 77–8, 78n.13, 82–3, 85, 95, 100, 106, 110, 185
prisoners of war (POWs) 51, 181
proportionality
contractual 112, 114–15
jus ad bellum proportionality
see: jus ad bellum-> proportionality
jus in bello proportionality
see: jus in bello-> proportionality
jus ex bello proportionality
see: jus ex bello-> proportionality
pre-contractual 114, 148
"purism" 14n.15
see also: Individualism

Quinn, Warren S. 26n.55

Raz, Joseph 115n.36, 194
exclusionary reasons 59n.39
legitimate authority 52–3, 85, 106, 124, 189–91
Rawls, John 68–9, 72, 174n.22
duty of nonintervention 96n.39
theory of justice 53–5
veil of ignorance 44
world government 92
reciprocity 3nn.9,11, 167

Reductionism 10
see also: Individualism; Revisionism
remedial rights 8, 156, 163–4, 166, 174–5
responsibility (as a condition to make one liable to defensive killing) 21–2, 24, 29–31, 33–5, 129, 136–7, 151–2, 160, 173–4
reprisal 166–9
see also: retaliation
retaliation 164, 167, 167n.6, 168–71, 176, 179, 188
see also: reprisal
revisionism 5–7, 9–10, 13, 15–16, 18, 22, 28, 33–4, 36n.75, 113, 116, 121–4, 175, 180, 190, 192, 194–9
see also: Individualism; Reductionist approach
'riddle of war' 102
right-based morality 172, 197
Ripstein, Arthur 2n.3
Rodin, David 16, 18, 21, 195n.7
Global state 92–5, 196n.8
'role ethics' 61n.41
robots 139–40
Ryan, Cheyney 12n.10, 116n.1

Scanlon, Thomas
contractualism 3n.5, 40, 66–9
secession 87
self-determination 89–92, 95–7, 153, 169–70, 182–4
self-defense 7–8, 15–17, 21–4, 29, 35, 77, 92–3, 96n.39, 99, 114, 116–18, 121–2, 140–1, 143–9, 161, 165, 173–4, 176–8, 178n.28, 182, 184–8, 190–2, 200
'collective self-defense' 15
see also: Article 51; self-help
self-help 4, 76, 92–3, 96, 98, 116–20, 145, 182, 185–6
Shue, Henry 134
Simmons, John 48n.23
social rules
Hart's theory of
see: Hart, H.L.A.
social roles 20–1, 68, 117, 124, 151–2
social contract tradition 68–9, 181
sovereignty 19–20, 71–2, 112–13, 120, 171–2
stateless nations 7–8, 83–4, 89–91, 96–7, 119–20, 140, 152–3, 156–7, 180, 182, 184, 187–8, 191
see also: non-state actors
state of nature 7–8, 39, 69, 89, 137, 162, 179, 189, 191
statism 180
Suárez, Francisco 2n.4
supreme emergency exemption 171–9, 189
symmetry thesis 13

Tadros, Victor 52n.27, 62n.44, 130n.26
targeted killing x

terrorism 89n.28, 154n.43, 157
Thomson, Judith J. 161
  on the relevance of intentions to permissibility 26
  trolley problem 26
treachery 129–30
trolley problem 24

uncompromising wars 99–101, 103–7, 115, 185
uniform
  the duty to wear a uniform 153, 188
United Nation Charter 1–4, 7–9, 10n.5, 16–18, 20–1, 71–2, 74, 77–8, 80–1, 83–6, 92, 95–6, 103, 106, 193
  Article 2(4) of the 4, 11, 73, 119–20, 156, 170, 182
  Article 51 of the 4, 7–8, 11, 16, 20, 73, 88–9, 93, 119–20, 156, 182
utilitarianism 67n.53, 173, 182

Valentini, Laura 52n.27
veil of ignorance 44, 68–9
Vitoria, Francisco de 1n.1, 2n.4

Waldron, Jeremy 3
Walzer, Michael 7, 9–10, 28, 82–3, 179n.30, 187
  contractrian arguments 3, 197
  Independence Thesis 10–11
  threats to territorial integrity or political independence as just cause 11–12
  legality of the intentional killing of combatants 12–13
  legality of incidental killing of civilians 12–13
  on civilian immunity 133–4
  on Jeff McMahan 28
  on moral equality of combatants 117
  on reprisals 167n.7
  supreme emergency exemption 171–3, 176–7
weapons of mass destruction (WMD) 8, 105n.19, 164, 174–7, 189
  see also: nuclear weapons
Whitman, James 102–3, 107–9, 109n.27
Williams, Bernard 131
world government 92

Zohar, Noam 9n.4